The First ADIRONDACKERS

The First ADIRONDACKERS

12,000 Years of Indigenous Peoples in the Adirondack Uplands

Curt Stager and
David Kanietakeron Fadden

NORTH COUNTRY BOOKS
Essex, Connecticut

NORTH COUNTRY BOOKS
An imprint of The Globe Pequot Publishing Group, Inc.
64 South Main Street
Essex, CT 06426
www.globepequot.com

British Library Cataloguing in Publication Information available

Library of Congress Cataloging-in-Publication Data available
ISBN 978-1-4930-8979-6 (paper : alk. paper)
ISBN 978-1-4930-8980-2 (electronic)

∞™ The paper used in this publication meets the minimum requirements of American National Standard for Information Sciences—Permanence of Paper for Printed Library Materials, ANSI/NISO Z39.48-1992

To our parents.

Given the abundant proof of activity and occupancy in various forms, why does the myth of an empty space with no precolonial history persist in the Adirondacks?

—Melissa Otis, *Rural Indigenousness*

CONTENTS

OHENTON KARIHWATEHKWEN
(oh-HON-doh gah-lee-wah-DEH-kwon)

"Words Before All Else"

For centuries, a traditional thanksgiving address has been spoken by the Haudenosaunee, also known as the Iroquois or Six Nations, at the beginnings of important gatherings. The intent of the address, the "Words Before All Else," is to bring people's minds together as one by greeting and thanking the natural world together. Normally, a speaker is chosen to narrate the address, and a response of agreement is voiced by those listening at the end of each section. Let us begin this time together with you, our readers, with a written form of it.

The exact form of Words Before All Else varies from person to person, but it always expresses gratitude and instills a sense of belonging to the earth, our home. This version of it has been crafted by me, David Kanietakeron Fadden, following the guidance of traditional mentors as well as my own experiences as a storyteller and interpreter of Haudenosaunee history and culture.

The Words Before All Else address serves here as a welcoming bridge between cultures that helps us recognize our connections to one another and to the natural world. We are each a small part of the larger family of Creation, and every part of that Creation is best considered as a family member. The earth is our mother. The sun is our elder brother. The moon is our grandmother. We must treat these beings with respect as we would treat members of our family, not as mere resources but as relatives.

Now, with sincere hearts and united voices, let us send our greetings and thanks to each part of Creation.

Onkwehshon:'a
(ohn-gwe-SOON ah)
"The People"

We recognize here the continuing cycles of life to which we all belong and see that we are living in connection with each other and all things. We now bring our voices together and thank one another as people. Now our minds are one.

Iethi'nistenha Ohontsia
(yeh-tee-nee-STAH-hah oh-HOON-jah)
"Mother Earth"

We are thankful to our mother, the earth, who gives us all that we need for life. She supports our feet as we walk about upon her. It gives us joy to know that she continues to care for us as she has from the beginning of time. To our mother we send greetings and thanks. Now our minds are one.

Ohneka'shon:'a
(oh-neh-gah-SOON ah)
"The Waters"

We turn our attention to the waters of the earth. They quench our thirst and wash our bodies. We would not exist if not for the gifts of the waters. They come to us in many ways. We see and feel the water in streams, rivers, ponds, lakes, rain, and oceans. With one mind and voice we send greetings and thanks to all water. Now our minds are one.

Kentsionshon:'a
(gon-jun-SOON ah)
"The Fishes"

We now look to the beings who live in water, the fishes. They continue to obey the Creator's instructions to cleanse the water. They provide food and nourishment. We send our voices to the fishes with words of thanks and greeting. Now our minds are one.

Tsi Shonkwaientho:wi
(jee soon-gwa-yon-TOH wee)
"The Vegetation"

We turn our eyes and hearts to the vegetation of the earth. Flowering plants bring beauty to the world with their variety of color and fragrance. They help many life-forms as food and medicine. We join our words together to send greetings and thanks to the world's bountiful vegetation. Now our minds are one.

Kaienthohshera
(gah-yon-toh-SEH-lah)
"The Food Plants"

We look now to the food plants of the world. From seed to fruit, they continue to provide the nutrition our bodies need. The Three Sisters—maize, beans, and squash—grow tall and healthy every year, and the berries taste sweet. Because of this we, with one mind, send greetings and thanks to the food plants. Now our minds are one.

Konti:rio
(gon-DEH-lee-yoh)
"The Animals"

We now look to the animals who live all around us. They provide us with meat, clothing, and tools. They send us lessons about the environment we live in. They remind us that we live together with them as cousins on the earth. We are grateful for their continued existence and send them greetings. Now our minds are one.

Okwire'shon:'a
(oh-gwee-leh-SOON ah)
"The Trees"

We turn our thoughts and gaze to the trees. From the tallest to the smallest, the trees provide us with many gifts. They provide shelter and warmth during the colder months. They give us sweet sap, fruit, and shade during the warmer months. They sing to us as the winds pass through their branches. Trees are symbols of peace, unity, and strength. With one mind, we send greetings and thanks to the trees. Now our minds are one.

Otsi'ten'okon:'a
(oh-jee-den-oh-GOH ah)
"The Birds"

We now look up to see the birds flying around us. We can hear their voices from the forest and near our homes. Their songs bring pleasure to all who hear them. They remind us to appreciate life. They vary in size and color, and each is a beauty to see with the eagle as their leader and guardian. We greet the feathered beings and send thanks. Now our minds are one.

Owera'shon:'a
(oh-weh-lah-SOON ah)
"The Four Winds"

We send our thoughts to the four winds. They bring changes in the seasons. They bring the rain, warmth, and cold. The air is renewed with their presence. Without the wind, life would be stagnant and still. They come from four directions, and each brings renewal and freshness. With one mind, we thank the winds. Now our minds are one.

Ratiwe:ras
(lah-dee-WEH lahs)
"The Thunderers"

We now listen for our grandfathers, the Thunderers. They come from the west bringing loud voices and power. They quell the negative aspects of life and demand attention when they arrive. They bring rain and help replenish the rivers and streams. We send greetings and thanks to the Thunderers, our grandfathers. Now our minds are one.

Kionhkehnehkha Karahkwa
(kyoh-KEH-neh-gwah gah-LAH-kwah)
"The Sun"

We now look to the sun, our elder brother. He appears from the east each day to bring the light and warmth that continue to bring life to the earth. He enjoys the heat and competition of sport, especially lacrosse. He is the source of fire that warms us. We now send our words of greetings and thanks to Elder Brother, the sun. Now our minds are one.

Ahsonthenhnehkha Karahkwa
(ah-soon-toh-NEH-hah-gah gah-LAH-kwah)
"Grandmother Moon"

We look to the night sky and see our grandmother, the moon. She lights the night sky and watches over female life on earth. She moves oceans and helps us determine the seasons and measure time. She oversees and guides the birth of children. Her gentleness brings calm and tranquility to those who see her. With one voice, we send greetings and thanks to Grandmother, the moon. Now our minds are one.

Otsistanohkwa'shon:'a
(oh-jees-dah-noh-gwa-SOON ah)
"The Stars"

We turn to the countless stars that spread across the night sky. They guide us as we travel at night and bring dew to our gardens and plant life. The Milky Way is our path lined with strawberries when we pass from this world. Stories we tell about the stars teach our children about life. With one voice, we give thanks to the stars of the night sky. Now our minds are one.

Karihwanho:ton
(gah-lee-wah-HOH doon)
"Closing"

We have come to the time when we close these words. We send our thanks to those whom we might have left out. It was not our intention to forget them. Those who remember can send their greetings in their own words. Now our minds are one.

INTRODUCTION

The First Adirondackers is a book with a mission. Its main purpose is to challenge a widespread, long-standing belief that the Adirondack uplands of northern New York were uninhabited before Europeans arrived. Although local Indigenous traditions hold that people have called this region home since time immemorial, most non-Indigenous authors to date have more or less denied the very existence of early Adirondackers. As a result, many residents and visitors alike assume that people are newcomers to this wild and beautiful landscape.

Nothing could be further from the truth. By documenting and describing the long legacy of human presence here, *The First Adirondackers* not only helps redefine what it means to be an Adirondack resident; it also contributes to a more complete understanding of America itself.

In his book *Rediscovering America*, the Western Shoshone author Ned Blackhawk proposes that it is "impossible to understand the United States without understanding its Indigenous history." And yet the scarcity of Indigenous perspectives in historical narratives has long been a problem throughout the nation. How else, for example, could so many of us mistake the powerful Indigenous influences on our lives for foreign imports? "Italian" tomato sauce, "Irish" potatoes, "Swiss" chocolate, and even the iconic modern snack, popcorn, all derive from crops that were developed thousands of years ago in what are now known as the Americas.

Blackhawk has written that "it is time to reimagine U.S. history outside the tropes that have bred exclusion and misunderstanding" by replacing weak interpretive tools of the past with better ones. To help further that goal, we will explore on these pages the interweaving of environment and cultures in the human tapestry of the region we know best, the six-million-acre Adirondack State Park. By challenging predominant Eurocentric narratives here, we are not erasing history, as some ill-informed critics might propose. Rather, we are helping to recover a proud and wonderful history that has too often been erased.

Like the vast mosaic of public and private lands that it describes, *The First Adirondackers* is a cultural crossroads in which multiple worldviews meet, both figuratively and in the personal lives of my coauthor, David Kanietakeron Fadden, and myself. Together, we use our different backgrounds to help address outdated ideas of who belongs here in the uplands, a region that we define as everything situated between 1,000 feet elevation and the top of Mount Marcy, a mile above sea level.

To further illustrate what we are up against, consider this portion of a recent podcast in which the guest and host deny any significant Indigenous presence in the uplands. It has been paraphrased slightly for clarity and brevity, but the main content is intact.

> GUEST: *"Before the 1700s there was nobody living in what is now the Adirondack Park. There was no reason for people to want to live there. The Adirondacks have a short growing season and poor soil, so Native Americans only passed through the uplands so as to leave places that were better than the Adirondacks and reach some other place that was also better. People didn't enter the interior until the nineteenth century or so."*
>
> HOST: *"I learned at an early age that Native Americans didn't live up here and just passed through it because conditions in the Adirondacks were too brutally harsh and cold. Some people argue with that idea, but it's common knowledge."*
>
> GUEST: *"There have been claims by people who don't know what they're talking about that the first settlers were Native Americans, but it's not true. They probably do it for purely emotional reasons, but no history supports their claims. This information I'm giving you here is the kind I hope people will still be using a century from now."*

We share those comments here not because they are unusual but because, unfortunately, they and similarly misguided statements are commonplace. They arise from more than a century's worth of books and articles that misrepresent the deep Indigenous heritage in the uplands or reject it entirely.

Examples are everywhere, so ubiquitous that they are easily taken for truth. I reach into my knapsack for a tattered copy of *Guide to Adirondack Trails*, published by the Adirondack Mountain Club, and read that this landscape was "not hospitable to Native Americans, who used the region principally for hunting and occasionally for fighting." Reputable experts whom you will meet in the following chapters have been pushing back against such accounts for years, and we offer this book in support of their ongoing efforts.

Let me now introduce my coauthor, present our credentials, and further explain how this book came to be. (Note to readers: a guide to the pronunciation of Indigenous names and terms is provided at the end of the book.)

David Kanietakeron Fadden is a member of the Wolf Clan of the Kanienkehaka, also known as the Mohawk Nation. Because his ancestral roots also extend into Wabanaki and European cultures, he is a living embodiment of the diverse heritage we explore in this book. Dave has recently taken the helm of the Six Nations Iroquois Cultural Center in Onchiota, a small rural community in the northern Adirondacks, which was created in 1954 by his late grandfather, Ray Tehanetorens Fadden, and grandmother, Christine Skawennati Fadden. Dave now represents the third generation of his family to run it. Among his many talents is the building of bridges between cultures through storytelling and the visual arts in ways that are welcoming, creative, and informative.

Like his Indigenous ancestors, Dave lives in both the uplands and lowlands, sharing his time between Onchiota and the community of Akwesasne in the Saint Lawrence Valley. Carrying forward the work of his grandparents and his late father, John Kahionhes Fadden, Dave is a caretaker of ancient traditions that link him to all life on earth. He is a master storyteller and also a renowned painter who has pioneered new ways of blending cultures in visual forms that communicate the humanity and vitality of Indigenous peoples.

My background is that of a professor of natural sciences at Paul Smith's College in the northern Adirondacks whose worldview, like Dave's, links me to all life on earth. Most of my research involves reading long environmental histories in layered deposits of ancient sediment recovered from the floors of lakes around the world, but my other scientific interests range from coral reefs to climate change and human history. In addition to formal college-level teaching, I've also learned to engage informal audiences through writing, radio, and film and, like Dave, I enjoy building intercultural bridges through the arts.

The evidence for a deep human presence in the Adirondacks that we will share here has been gathered by visiting archaeological excavations and museums, interviewing experts, reading extensively, and documenting locations in the uplands where ancient items of Indigenous origin have been found. The map shown on page xix registers more than three dozen such sites.

Apart from the ongoing need to challenge the absence-fable in print, an additional inspiration for this book was a project that Dave and I participated in recently, a science-art collaboration funded by the Lake Champlain Basin Program. The project was intended to raise public awareness of watersheds and their influence on water quality in Lake Champlain, which lies between the Adirondacks and the Green Mountains of Vermont. What made the approach most unique was the connectivity and cultural diversity of the physical setting and of our team, which included both science-friendly artists and arts-friendly scientists from a variety of ethnic backgrounds.

My role in the project was to explain in scientific terms how people are elementally linked to the land, life-forms, and waters of the Champlain-Adirondack region. Several Indigenous artists from the North Country then interpreted that information through their favored media and cultural perspectives. For his role, Dave painted my technical descriptions of chemical elements and microbes into the background of *Water is Life*,

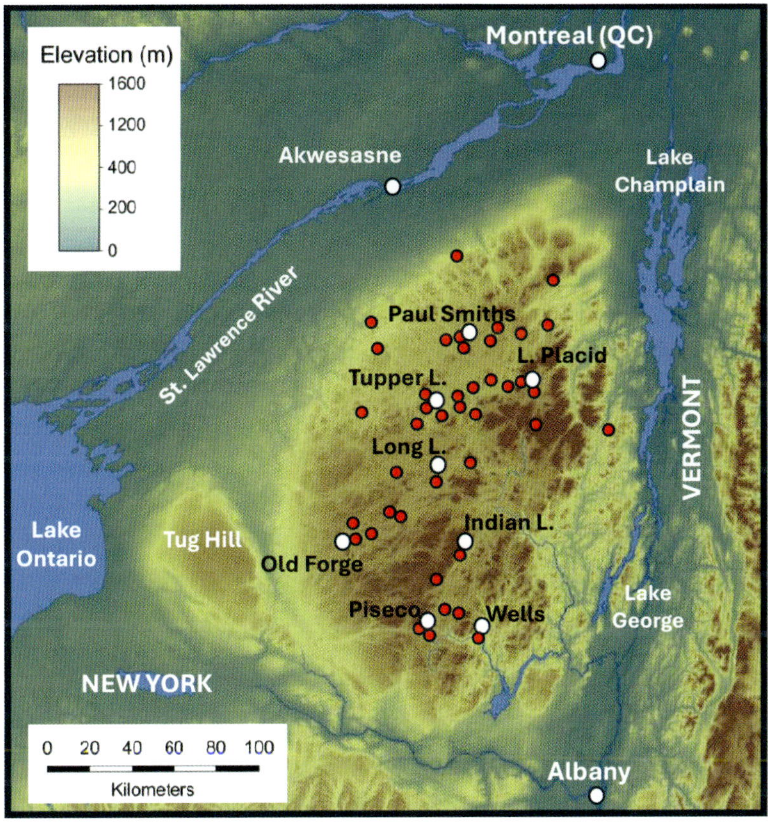

Site map showing the Adirondack uplands (golden to dark brown), selected communities (white dots), and approximate locations of some of the sites where ancient items have been found (red dots). BASE MAP COURTESY OF DAVID FRANZI.

in which an Indigenous woman cups the essence of a watershed in her hands. Close examination turns the vegetation around her into a vibrant mosaic of traditional iconography, faces, molecules, various living beings, and humorous flights of fancy, a blend much like the land and peoples of the Adirondacks themselves. His skillful artistic and spiritual interpretations of the science infused it with raw emotional power of the sort that we hope you will sense in these pages, too.

The First Adirondackers is not a comprehensive historical document. We will leave that important work to other investigators, and we have relegated most technical details and citations to the Chapter Notes and References sections at the end of the book. Nor does it fully represent the ethnic and professional communities to which we belong. Rather, it is a way for Dave and me to combine our personal experiences and perspectives in written and visual form to push back against the false claim of absence. To accomplish that, we will employ the same sort of cross-cultural alliance between science and art, mind and heart that infused the watershed project.

There is a Haudenosaunee tradition of describing such collaborations between worlds with a metaphor, that of traveling along a river together but in separate watercraft. One artistic representation of the shared journey is a ceremonial belt of cylindrical clam-shell beads, known as "Two-Row Wampum," which depicts two parallel tracks in purple on a background of white. Images of such a belt serve as section markers throughout the text as a reminder that this book is a shared journey.

In that spirit, Dave's version of the Haudenosaunee thanksgiving address precedes this introduction and his traditional account of the origin of the world precedes my science-based account of the origin of the Adirondacks. Both of us present our sections of the book with the full conviction of our own cultural perspectives, but we do so in respectful tandem with one another as fellow Adirondackers. Most of our references to Indigenous cultures employ their traditional names (e.g., Haudenosaunee rather than Iroquois, Kanienkehaka rather than Mohawk), though some exceptions are made for the sake of clarity. We also acknowledge the kinship of all living beings that is evidenced by Indigenous and scientific knowledge alike by referring to nonhumans with terms like "she" or "who" rather than the impersonal "it" or "that."

The following chapters are linked to one another by the river of time. They will lead us on a journey through history that is chronicled step by step through imaginary vignettes on a beautiful lakeshore knoll where my place of employment, Paul Smith's College, is located.

The knoll overlooking Lower Saint Regis Lake, with Saint Regis Mountain on the horizon.

As the millennia pass we will encounter caribou-hunters at work 12,000 years ago, a young woman being tempted to collect a forbidden stone implement 8,000 years ago, an ice-fishing excursion 5,000 years ago, an encounter between Adirondackers of differing cultural backgrounds 600 years ago, and the moment in 1858 when Indigenous residents were dispossessed of the knoll. The vignettes are chronicled in the present tense to help bring the past to life, and the characters featured in them are portrayed as realistically and respectfully as possible through my words and Dave's illustrations. Finally, Dave concludes our journey with his personal recollections of troubled times in the North Country as well as a vibrant renaissance of traditional culture that is unfolding today.

Woven into the narrative are scientific insights into the environmental history of the region and evidence for the deep human presence here along with relevant photographs (my own, unless otherwise indicated). We also include a reexamination of place-names in the region. Some readers might be surprised to learn, for example, that the Haudenosaunee-derived term "Adirondack" did not traditionally refer to the mountains at all. Rather, it was imposed on the region by a Euro-American surveyor in a form that, technically speaking, is pronounced incorrectly.

In addition, we retrace the story of how most of the upland region passed from Haudenosaunee to non-Indigenous hands in recent centuries, a complex tale whose details are unfamiliar to most Adirondackers. With the aid of reputable historians, we show that the three largest transfers occurred around the time of the American Revolution, with varying degrees of fairness or legality. But we go deeper than that, too. The long-term perspective employed in this book reminds readers that the very first inhabitants of the Adirondacks were not Haudenosaunee, either, but other Indigenous peoples with even older connections to the uplands.

Following the main text, we include an appendix that can serve as a practical training manual for challenging the false claim of absence with the aid of information presented in this book. As with an anatomical appendix, such misleading claims are unwholesome dead ends. We hope that our rigorous treatment of this one can help readers deal with similar cases of "historical appendicitis" they might encounter on their own.

The timescales we will be working with here are easier to write and read about than to grasp on a gut level. After all, most of us never live out a single century, but we nonetheless consider a full lifetime to be long, not to mention the parts of it that can seem to drag endlessly on occasion. It takes practice to fully embrace the vastness of this tale, which can make it difficult to recognize just how much is missing from what has often passed for Adirondack history.

The following exercise can help forge such a link between thought and feeling. Imagine a human lifetime being represented by the small stick figure of a human. For the sake of easy math, let's also consider an average lifetime to last fifty years, taking into account the various hazards faced by people through the ages. Using such fifty-year stick figures to represent sequential lifetimes, we need to link only five of them like wampum beads to span the entire existence of the United States. Ten of them take us back to the arrival of Columbus five centuries ago, an event that many Americans consider to be the starting point of the American saga. We can encircle them all in a small box here.

But those ten figures take us only a short way into the real story of people on this landscape. Roughly twice as many are needed to cover the period during which intensive horticulture was widely practiced in the North Country. We can add this longer sequence and encircle both together to represent the last millennium.

But even the horticultural period is still only relatively recent history. Most of the full saga had nothing to do with the kinds of cultivation-dependence that misguided sources say prevented people from living in the Adirondacks until modern times. To represent the complete sequence of human lives we will need something more like . . . well, let me show you.

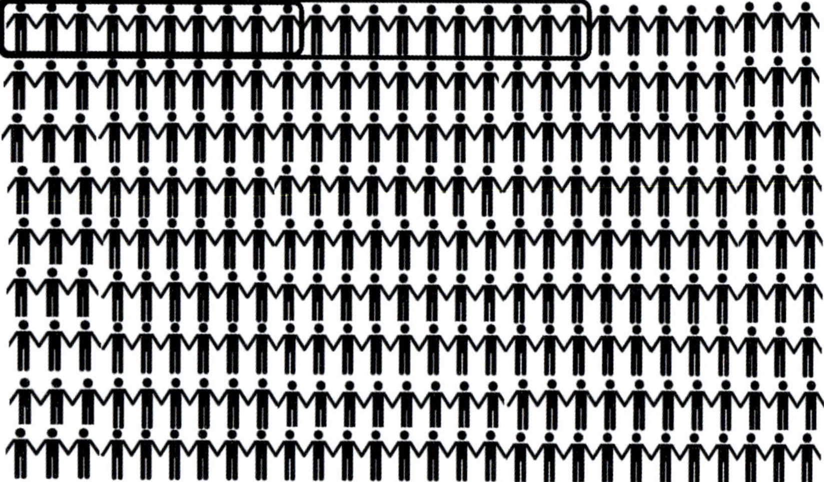

Each of these stick figures represents real people, more like an entire generation in fact, all of whom lived out their lives here. Although some scholars have referred to the original inhabitants of this region with crude terms like "early man" or "primitive man," do not be misled. They were all Adirondackers, every bit as genetically, magnificently, and sometimes maddeningly human as anyone alive today. They had names, dreams that were realized or never fulfilled, unique personal triumphs and tragedies, births and deaths. Most of them will forever remain invisible to written history, but their presence is still felt by descendants who call these uplands home.

This is the deep, rich tapestry of life in the Adirondacks that Dave and I will explore with you in *The First Adirondackers*. We hope thereby to honor those long-ago people by bringing them more fully into the realm of the remembered, and also to better understand and appreciate our own place on this numinous landscape that shapes and sustains us today.

Knowledge alone is sterile and action is blind until the mind travels together with the heart on shared rivers of understanding and inspiration. Therein lies truth, magic, and our best hope for a better future.

Shé:kon, and welcome to the Adirondacks. The *real* Adirondacks.

ORIGINS

All people have stories of where they come from and why they are here. Some traditional stories tell of human emergence through reeds from an underworld existence, or of a time when all was one and covered with smoke. Others describe an entity who created the heavens and earth by his will. Some cultures have no formal explanation of Creation at all and simply live in the present. There are common threads among many of these traditions, including a fundamental lesson of being good people, but there are differences, too. Views of, and relationships with, the natural world can range from humans having dominance over everything on Earth to addressing the natural world as family.

The Haudenosaunee account of Creation is one such tradition that provides guidance for living our lives and our perceptions of how the world works. As with any oral tradition, aspects of the account change from narrator to narrator and over the generations. However, its main lessons remain the same. The abbreviated version of the Haudenosaunee Creation account that I present here highlights only a few of the more important elements of the story as told by my grandfather, Ray Tehanetorens Fadden, and other members of the Haudenosaunee nation.

The story begins with a time when the earth was dark and water covered everything, everywhere . . .

The only inhabitants of the earth in the earliest times were those who thrived in water. The beaver, otter, and muskrat found it pleasant and lived there comfortably. The ducks, loons, and geese felt at ease in that watery world, too.

High above, there was also a place called the Sky World where there lived beings who closely resembled what we look like today. The Sky Beings had a peaceful and contented existence. Every breath of air in the Sky World was sweet and fresh, and the wind was not too cold or too violent. The land had rolling hills and streams that flowed throughout the countryside. The light was not too bright and not too dark. In short, this place was a paradise.

One of the roles of the Sky Beings was to take great care of a tall tree in the center of the main village and to make sure that no harm ever came to it. This tree had many types of fruit, each with a different taste, smell, and color. The Sky Beings were instructed not to pluck fruit from the tree. Rather, if they wished to eat the fruit, they were supposed to wait for it to drop to the ground. Light emanated from the great tree's flowers and sent a soft glow throughout the Sky World for all to see.

In the main village there lived a woman who was expecting a child. She, like many women who are with child, had cravings for many kinds of food at different times. Her husband, the caretaker of the celestial tree, spent all hours of the day searching for and gathering various foods to bring home to his wife. Day after day, week after week, and month after month the husband searched the Sky World for everything that was edible, but his wife's yearning was still not satisfied. She became agitated by this craving that continuously growled in her stomach.

One day, the woman sat at the base of the sacred tree wondering what kind of food could satisfy her. She looked at the tree and realized that she had tasted every type of fruit that grew on it. She had even gnawed on small branches that fell to the ground. It was then that she noticed the trunk of the tree and how its roots grew down into the soil. She wondered what those roots might taste like, assuming that they would be delicious as they carried nutrition to the whole tree. She dug around in the soil to find a small root, but with little success because the roots were firmly embedded.

Her husband neared and noticed that she was digging at the dirt around the tree. He hurried to her and firmly reminded her of the tree's importance and the rule that no one was to harm it. She argued that the tree was strong and asked how taking one small piece of root could damage such a big strong tree. Her husband used his best words to try to make her understand the sacredness of the tree. For a long time he spoke to her, but her voice rose to the point of desperation. Even her stomach seemed to call out.

After much discussion, the husband decided that perhaps removing one tiny piece of root would not harm the tree. Reluctantly, he dug around the base of the trunk and found a small root that looked insignificant. He tugged at it, gently at first, in hopes that it would break free. But after struggling for some time to break the root, he tugged with all his might. To the astonishment of both husband and wife, the sacred tree creaked. Before they could do anything else it tilted, fell, and landed with a tremendous crash.

The entire village rushed to see what had happened. As they gathered around the fallen tree, the husband and wife looked to where it had stood and saw a large hole there. Peering into the abyss, they saw a vast, rippling water far below with birds gliding over it.

The wife, or Sky Woman as she is referred to now, is said to have looked too closely into the hole. She lost her footing and stumbled into it. She grasped the sides of the hole and tried to hold on and pull herself back up, but as she struggled the roots and plants she was holding on to gave way. She fell down through the opening with tobacco and strawberry plants clutched in her hands.

At first, the waterfowl below were frightened when they saw a hole in the sky and a Sky Being falling toward them. Then they noticed that she could not fly and that her hands and feet were not webbed so she would not be able to swim well. They became concerned and flew closer to see what they could do to help her.

The geese flew near enough to slow Sky Woman's rapid descent by letting her rest on their backs. As she neared the great water, a large turtle surfaced and offered his shell as a place to set her down on. The geese flew to the turtle's back and Sky Woman set her feet safely upon his shell.

As Sky Woman looked around, all she could see was water with no sign of land. Being curious, the otter, beaver, and muskrat swam around the turtle to get a glimpse of her. She saw them and asked if they knew of any other place where she could live and walk around. The water animals swam to her and told her that there was no such place as far as they could see or swim. However, they said that mud existed deep down under the great water.

Sky Woman asked if any of them could swim deep enough to gather some of the mud for her. The beaver offered to try and dove swiftly down into the depths of the water. Time passed, and then his limp body floated slowly to the surface. He had failed.

The otter offered to help, too. With barely a ripple he dove straight into the dark waters below. More time passed, longer than that taken by the beaver, but then the otter's limp body also floated to the surface. He, too, had failed.

Finally, the muskrat took his turn. He dove with determination, and after a time longer than that of the beaver and otter his body floated to the surface, as well. Saddened by the loss of these Water Beings, Sky Woman went to the body of the muskrat to show her gratitude. Then she noticed something clutched in his tiny paws. It was mud from the bottom.

Sky Woman thanked the beaver, otter, and muskrat and took the mud from the muskrat's paws. She leaned down and spread the mud onto the turtle's shell. Seeds from the tobacco and strawberry plants fell into the mud and began to grow. Sky Woman sang a song from the upper world and shuffled her feet as she danced counterclockwise. Her feet pushed the seeds into the mud and the sacredness of her song encouraged plants to sprout.

Turtle's back shifted and danced with the song and began to grow in size. Soon the great turtle's shell grew so large that Sky Woman could no longer see the edges of it, and she began to walk around on the enormous, soil-covered shell.

In time, Sky Woman felt a baby moving inside her, and soon a girl was born in this new world. The baby and her mother lived contentedly as the turtle's shell continued to grow. However, as the young girl became a young woman, she became restless and lonesome. She wandered Turtle Island and explored her world as each day passed.

One day, she wandered too far from her home and became lost. A spirit, West Wind, saw her and swirled down to her. She found herself becoming dizzy as the wind whipped all around and picked up dust and debris, causing her to faint. When she came to, the wind had subsided and she saw that two arrows had been placed on her belly. She returned home to her mother and told her what had happened. Sky Woman told her daughter that she was going to give birth to twins and that the father was a powerful spirit.

As time passed, the daughter's belly began to swell. As the babies in her womb grew, she could feel and hear them communicating with one another. At times it seemed as if they were arguing and wrestling inside her. Two twin boys were preparing to be born, and she wondered how this would come to pass.

One twin was very calm and patient. He knew that the time of their birth would come eventually, and that he and his brother would know when. The other twin was unsettled and anxious. He wanted to be set free quickly.

As the day of their birth drew near, the calm twin counseled his brother to wait until it was time. His brother, however, was becoming more and more distressed, eager to find his own way out. He noticed a faint light shining through the wall of his mother's womb. With his skin sharp as flint, he scraped at the wall of the womb beside him and eventually made an opening under his mother's arm. The patient twin had been born without incident in the natural way, but the impatient twin's method of birth caused his mother's death.

When Sky Woman came upon her newly born grandsons and their dead mother, she became distraught and angry. She didn't know who had caused her daughter to die. She became convinced that the calm twin had caused the death and decided to nurture and care for the anxious one instead. She cast the good-minded twin into the bushes where the beaver and other four-legged beings took care of him. His name became Sapling. Flint, his brother, lived under the care of his grandmother.

Sky Woman dug a hole in the soil and placed her daughter's body within it along with plants she had brought with her from the Sky World. From her daughter's body grew all of the food plants that anyone could need for survival. From her body grew the corn, beans, and squash that became known as the Three Sisters. From her heart grew sacred tobacco and from her feet sprouted strawberries and various medicinal plants. From that time to today, we refer to Sky Woman's daughter as our Mother Earth.

As time moved on, the twins grew and explored their world. They found that they had special powers to create living beings and move landforms, soil, and water. Sapling would blow his breath onto some soil and life would appear in the shape of beautiful animals like the deer, bear, wolf, and fox. He would mold the land to form smooth rolling hills and valleys full of life. He made the waters taste sweet and flow calmly through the valleys to settle in lakes and ponds for the fish and other Water Beings to thrive in. He created plants with flowers and fruits who smelled sweet and satisfied one's hunger.

Meanwhile, Flint witnessed his brother's work and found that he could not make beings with such beauty. When he tried, snakes and venomous creatures emerged instead. Frustrated by his inability to create beauty, he instead manipulated his brother's work by placing sharp thorns on colorful plants and putting poison into certain fruits. He pushed the rolling hills together to make them jagged and sharp with steep drops and dangerous boulders that could tumble down at any time. He threw stones into the calm waters and twisted the paths of the rivers, making it dangerous to travel.

One day, Sapling heard a voice and saw a male figure standing before him. The man told him that he was the elder brother of his grandmother from the Sky World. He saw and liked what Sapling had created but also noticed that there were dangerous creatures around. Elder Brother promised that he would appear at the horizon each day and travel across the sky, bringing light and warmth to the earth, and that sunflowers would follow his path.

In time, Sky Woman became old and frail. Her days on Earth were coming to an end. When she died, Flint and Sapling argued about what to do with her body. They struggled with it until her head was flung into the night sky to become the moon, our grandmother. She still watches over all female life-forms, and babies still arrive according to her guidance. She will always light the night sky and guide the seasons.

As the first fruits ripened and animals multiplied, Sapling was ready to create humans. He gathered the finest living soil with the richest colors and added to it some of his own blood and some of his breath. Soon the first humans stood before him, alive and full of energy. He instructed the people to work together as one and create new life for coming generations.

He said that some of them may reach old age while others may not even survive long enough to walk the earth. Only he would know how many days a person would have for walking upon their Mother Earth.

The people now wandered the world and saw the plentiful gifts that Sapling had provided for them in the form of animals, water, and plants. Meanwhile, the jealousy that Flint felt toward his brother finally turned to anger and then to rage. He challenged Sapling to a battle. Flint used a cattail as a weapon, thinking that it would poison his brother. Sapling grasped a deer antler to defend himself with. Their mighty struggle caused the earth to shake. Mountains moved and lightning struck the ground.

Finally, Sapling defeated Flint and cast him away so that he would not interfere with his work any longer. He did not kill him, though. Instead, Flint was banished into the depths of the earth. Nonetheless, Flint's influence is still felt by humans in the form of anger, frustration, and mental anguish. He sends his evil messages to us on days when all seems to be wrong and not going in a peaceful direction. However, such days are temporary. Sapling's teachings and thoughts always prevail in time.

Sapling spoke with the humans and told them that his work was done and his time on Earth was at an end. He reminded them to appreciate the gifts that he had given them and to take care of them for the future. He told them that if they ever wished to talk with him, they were to burn sacred tobacco so the smoke could carry their words to him in the Sky World.

Sapling also promised to return to check on the well-being of his Creation. He reminded the people that their days were limited and that only he knew how long they had to live. He pointed to a trail that leads to the Sky World and told them that when their days were finished the trail would be lined with strawberries for them to eat as they traveled. That celestial trail is known today as the Milky Way. With that, Sapling finished his instructions and began his journey to the Sky World.

After a long time, Sapling reappeared on Earth only to find that everything had been neglected and the minds of the humans were divided. He gathered the people together again and instructed them to give thanks to all parts of Creation from the smallest blade of grass to Elder Brother the sun and everything in between. Those words of thanksgiving would help the earth flourish and grow as it was meant to do.

Sapling explained that resolving differences through such words without anger would allow humans to find happiness and contentment. Gatherings throughout the year would show gratitude for the gifts of the earth, and words of thanksgiving would be spoken at the beginning of each gathering to remind people of the shared gifts that support and nourish their existence on Earth. After delivering these teachings, Sapling departed again, having promised to return in the future if the lessons should become neglected or forgotten.

Using the instructions given to us a long time ago, we Haudenosaunee still say words of thanksgiving throughout the year. Some people use them every day. They remind us that our lives depend upon the health and well-being of all parts of Creation and our relationships with all life on Earth.

ICE, SUN, AND STONE

It is early autumn in the Adirondack uplands, and here in the twenty-first century we are standing at 1,600 feet elevation atop a thinly wooded knoll overlooking Lower Saint Regis Lake. Our sun-warmed faces feel the soft press of a breeze that has crossed the water from the forest-cloaked slopes of Saint Regis Mountain on the western horizon. It also coaxes a sweet, resinous fragrance from the bristling needles of red pines who whisper overhead. To the right of us lies Paul Smith's College, whose lakeshore campus was a renowned luxury resort in the wilderness during the late 1800s. However, there is little evidence of people in front of us apart from a log lean-to on the far shore and a fire tower atop the mountain.

We have this spot to ourselves, and our imaginations are free to wander. Let's now turn the knoll into a personal time portal that can transport us deep into the past, back to when the Adirondacks first emerged from beneath the last great ice sheet.

Thousands of years pass in reverse until we reach another autumn day on the knoll. The mountain and lake are still here, but the differences we see around us are shocking. All of the trees, shrubs, and herbs have vanished, and the mountain stands naked on the horizon, clad only in random patches of lingering snow. Every trace of vegetation has been replaced by sterile heaps of sand and gravel atop smooth gray bedrock.

As we look to our right we see the future site of Paul Smith's College, minus the college. On this day thousands of years in the past, deglaciation is well under way and a barren outwash plain of grit and gravel lies where we expected to see the campus. Cold, glistening streams that intertwine atop it are draining silty meltwater into the lake and turning it milky with fine powder that the ice sheet has scoured and scraped loose from pulverized bedrock. A sinuous esker ridge that was once the sandy bed of a river beneath the ice sheet coils among stranded boulders and other glacial debris like a sleeping serpent. And over on the northern horizon is the source of it all, a mile-high ramp of melting ice. The edge of it slopes gently upward to where the main body of the dying ice sheet still crushes eastern Canada.

Here amid this scene of devastation and rebirth, our story of people in the Adirondacks is about to begin. In the chapters to come, we will return to this same lakeside knoll repeatedly to watch the passing ages carry us forward to our own current chapter of the saga. But first, let's reach even deeper into the past.

The geological history of the Adirondacks sets a grand stage upon which the human story has played out, and it echoes much of the Haudenosaunee Creation account. It even begins with a watery world and includes a huge, rocky mound shaped like the shell of a turtle who has surfaced from the depths.

Over a billion years ago, an ocean covered the place where the Adirondacks now stand. Two continents slowly collided in the heart of it, one riding up onto the other to form a stupendous mountain range that rivaled the Himalayas in size. The deeply buried anorthosite and granite roots of those mountains would later become the lofty peaks of the Adirondacks.

So much time passed that raindrops and rivers wore those ancestral mountains down, flattening them pebble by pebble, grain by grain. The sea returned, depositing sand and mud that slowly hardened into layered stone. Those primeval marine rocks from half a billion years ago would later supply Indigenous residents of the North Country with tool-grade quartzite and flinty chert.

Roughly ten to twenty million years before our current century, churning currents of magma deep beneath the earth's crust hoisted the old roots of the former mountains up through the ocean deposits. A vast, rounded dome rose like the shell of a great snapping turtle amid the remnants of the ancient sea. The tallest peaks on the dome now reach as much as a mile above sea level and are still rising by several millimeters per year.

The geological creation story of the Adirondacks is still ongoing. The age-old struggle between uplift and erosion continues to this day, and the uplands sometimes shudder with earthquakes as the dome stretches and grows. During the last two million years or so, mile-thick ice sheets have repeatedly crept in from the north to bury the region for long millennia before halting and melting in place. Each ponderous wave of glacial ice scoured the uplands and smeared more sand, silt, and gravel over the stone turtle's shell. After the latest push heaped debris into the shapes of Long Island and Cape Cod around 20,000 years ago, the scarred surface of the uplands began to emerge again for good—or at least it can seem like that to short-lived beings such as us.

In fact, the ice age hasn't ended yet. The current warm phase during which major civilizations emerged worldwide is only one of more than a dozen temporary interglacial reprieves in a long-term trend of increasing cold. Although we are now also overheating the world artificially with our fossil fuel emissions, it is simply a matter of ponderous geologic time before yet another ice sheet engulfs the northern half of this continent many thousands of years from now.

Most of today's Adirondack water bodies were born as the freshly exposed peaks and valleys cradled streams of meltwater atop thick beds of glacial debris. Where monstrous frozen chunks fell away from the ice sheet and melted, the pits they left behind in the sand and gravel became kettle holes and ponds. And where lakes of meltwater filled low areas at the foot of the shrinking ice sheet before draining off, they left behind silty deposits that would later become some of the richest soils in the region.

We know these things because their physical legacies are still with us. Consider, for instance, subtle patches of gray in the wave-cut rim of Bear Pond in the Saint Regis Canoe Area, where the canoe-carry trail to Little Long Pond begins. That's old glacial silt that had once turned the newly formed pond murky. Athletic rock climbers may often practice their moves on house-sized boulders near McKenzie Pond without recognizing them as ice-borne imports from elsewhere. And if you pull over beside a smooth rock face along the road between Meacham Lake and

Saint Regis Falls and gently run your fingers down it, you'll feel horizontal scratches that resemble marks from the teeth of a gigantic rasp. Those scars were ground into the hard bedrock by pebbles embedded in the base of the advancing ice sheet. Some of the fainter scratches in the mix might be even older, the ghostly echoes of previous ice advances that speak to us from more distant reaches of time.

Soon after the ice left the Adirondacks, the barren land began to brighten. Among the first pioneers were green, red, and golden splotches of hardy, encrusting lichens who made the boulders and cliffs look like a painter's palette. Next came spongy mosses, ground-hugging alpine flowers and grasses, and thick tangles of spruce. And shortly after the vegetation came, so too did the first animals.

However, before we consider the creatures who lived here, we need to establish when all of this took place. It is important to do so because so few remains of early animals have yet been found in the uplands that we can only guess what species lived here. And to guide that guesswork we need to establish a clear time frame, because many of the ice age creatures of what is now New York State became extinct soon after the ice sheet melted away. That leaves us with a relatively short period to align with the first chapter of associated human history in the uplands.

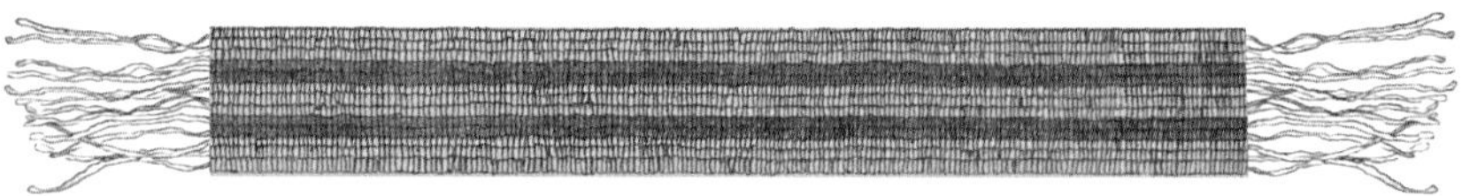

To determine when the last ice sheet left the uplands, we can use radiocarbon dating techniques on the remains of plants, animals, or microbes who lived here long ago. In the Adirondacks, lake sediments are among the most important sources of those datable materials.

During the summer of 1979, a graduate student from Brown University named Steve Jackson led a team of young assistants into the heart of the High Peaks in search of such sediments. Heavily laden with camping gear and sampling equipment, the researchers marched up to Lake Arnold, at 3,800 feet elevation one of the region's highest lakes. Over the next several days, Jackson and his team built a floating work platform from two inflatable rafts, anchored it in the center of the lake, and used long metal rods to force sampling tubes down into the soft mud of the lake bed.

Jackson recently recalled for me what happened when a passing hiker spotted them and reported their strange behavior to a ranger down below.

"There are some crazy people up there on Lake Arnold with a sailboat!"

Skeptical, the ranger hiked up the long, steep trail to check on them. What she found, of course, was not a mad sailing party but a landmark scientific exploration of the region's environmental history. The cores of sediment that Jackson's team retrieved from the bottom of the lake contained ancient pollen grains, conifer needles, and the remains of microscopic algae that reflected ecological conditions in and around Lake Arnold through the ages. They also yielded radiocarbon ages that reached back thousands of years.

By the end of the following summer, Jackson's team had sampled several other sites that ranged in elevation from about 2,200 feet at Heart Lake to 3,100 feet at Upper Wallface Pond. The bases of the oldest cores contained dense gray silt that had been left behind by the melting ice sheet. During the deglaciation, each freshly exposed lake had lost its cloudiness as silt-laden meltwater rivers receded along with the ice, and blankets of vegetation covered formerly loose glacial rock dust. By dating the brown organic lake mud just above the sterile gray zone, Jackson found that the High Peaks were mostly ice-free about 13,000 years ago, but that different sites yielded different dates of variable precision.

For example, the ranges of possible transition dates at each of the three lakes appeared to span anywhere from 600 to 1,600 years, making it difficult to pinpoint their exact timing relative to one another. Furthermore, there was no clear link between the date of ice retreat and elevation at those sites. Lake Arnold thawed out much later than low-lying Heart Lake, which might be expected from Arnold's higher, cooler location, but Upper Wallface at a similarly high elevation seemingly transitioned long before either of them.

During the winter of 2002 I told this story to my paleoecology students at Paul Smith's College and noted that more of this kind of work remained to be done with the aid of some improved analytical tools. By then a refinement of the radiocarbon dating process, called accelerator mass spectrometry, or AMS, allowed much smaller samples to be dated. Smaller sections of a core represent fewer sediment layers and therefore narrower time windows that help increase precision. Painstaking analyses of AMS dates on the annual growth rings within ancient tree trunks also permitted more accurate conversions of raw radiocarbon dates into so-called calibrated ages. For example, a radiocarbon age of 11,000 years converts to a true age closer to 13,000 calibrated years (all dates in this book are given in calibrated years). Perhaps, I suggested, the time was ripe for a new phase of investigation into the deglacial history of the uplands.

Two students, Jason and Sean, took the bait. They decided to follow Jackson's example by sampling Lost Pond, at 3,700 feet elevation in the High Peaks. Their goal was to answer the main unresolved question we had considered in class: Did the highest points emerge from the melting ice sheet first, or did the valleys open up first while ice lingered on the peaks?

After breaking trail for hours under a heavy load of coring gear in frigid temperatures, the students finally reached Lost Pond. They walked out onto its frozen surface, cleared a working space in the snow, and began drilling through the ice with a hand-driven auger to sample the lake bed below. Unfortunately, the ice was so thick that their auger was too short to penetrate it.

That didn't stop Jason and Sean. They returned to Lost Pond the following weekend. Better equipped this time, they collected a lovely core in which glacial silt lay beneath soft brown lake mud. The lowest layers of organic mud yielded a much-narrowed calibrated age range of 13,000 years, give or take a century or so. That was consistent with the date range from Upper Wallface Pond and notably older than the one from Heart Lake farther below. The hard-won weight of evidence from that student expedition now seemed to show that the tips of the tallest peaks emerged first as the ice sheet sank down around them—but the delayed transition to lake mud at Lake Arnold remained a mystery. Was it due to some error in the dating? Maybe so, but additional research in the uplands would suggest a more nuanced answer.

Other cores from sites with elevations slightly lower than that of Heart Lake, including Lower Saint Regis Lake, Moody Pond, Wolf Lake, and Brandreth Bog, have yielded a wide range of transition ages between 14,500 and 11,000 years, much like what was found in the High Peaks. Rather than following a simple relationship to altitude or vanishing all at once, it appears that big ice hung on longer in some places than others. That's not surprising, because the rugged landscape still forms distinct pockets of cold and warm microclimate today that could speed or slow a long thaw. For the sake of convenience, however, we'll say here that the uplands were largely ice-free by 13,000 years ago.

Sediment core studies such as these can also offer insights into what the local vegetation was like in those early days. Pollen grains and bits of foliage in cores from Brandreth Bog, Wolf Lake, Heart Lake, Lake Arnold, and Upper Wallface Pond have shown that tundra herbs, spruce thickets, and a scattering of pines spread across the uplands soon after the ice receded. The resultant botanical crazy quilt is known to ecologists as "boreal parkland," a habitat that covers much of Arctic Canada today.

In sum, the stage was set for animals to colonize newly exposed Adirondack landscapes roughly 13,000 years ago. With that time frame established, we can now guess what kinds of wildlife first populated the uplands, but we need not rely entirely on speculation in that regard. As we shall soon see, a remarkable discovery from a local gravel pit has given us some solid evidence to build upon.

One day in 1986, David Franzi, a geologist at SUNY Plattsburgh, learned of a surprising find in the Adirondack foothills near Elizabethtown. Embedded in a pocket of fine-grained lake sediment amid coarser glacial sand and gravel was a chunk of brown bone.

"It was a neck vertebra," he recalled when I asked him about it recently. "Not fully fossilized, so there was still plenty of carbon in it for dating."

Radiocarbon analysis determined that the animal bone was between 13,400 and 13,100 years old. A paleontologist concluded that the vertebra had belonged to a musk ox, a herd animal with heavy curved horns, a thick and woolly fur coat, and a geographic range that is now restricted to the far North.

"I'm guessing that this musk ox drowned in one of the meltwater lakes or rivers and sank to the bottom," Dave told me. Unlucky for the victim, but lucky for us.

The ancient musk ox bone shows that habitat suitable for large, plant-munching animals was available in the Adirondack region even before the uplands were fully ice-free. But let's address the elephant in the room, so to speak. We're talking about ice age mammals, after all. What about mammoths and mastodons?

The Northeast was once home to two kinds of cold-tolerant elephant: woolly mammoths and American mastodons. Their bones and tusks have often been found in Central New York and in the southern part of the state, but not in the Adirondacks. However, that doesn't necessarily mean they didn't live up here. Absence of evidence, after all, is not evidence of absence. Mammoths lived in open tundra and grasslands, while mastodons preferred spruce-rich forests, and both habitats were present in the Adirondacks shortly after the ice left. The reason for the lack of finds in the uplands thus far may simply be that the area is heavily wooded, thinly settled, and under-studied by paleontologists.

Bushwhacking through dense spruce stands would have been prickly going, but mastodons were well suited to it. Equipped with lumpy, wood-crushing molars and strong, stubby legs, they ate their way through boreal forests, as spruce twigs found inside some of the best-preserved carcasses attest. Mammoths had longer front legs and a more regal posture that was emphasized by a shoulder hump and curled tusks, and they mashed softer tundra herbs with flatter molars. Both species became extinct soon after the ice left the Adirondacks, but we can still infer their possible presence here during their waning days if they existed elsewhere in the Northeast then.

In 2012, scientists at the New York State Museum in Albany compiled a list of radiocarbon dates on mammoth and mastodon remains found in the state. The most recent mammoth age on the list was from Randolph, where a huge tooth yielded a calibrated age of about 12,000 years. The latest mastodon find thus far was from Ellenville, with an age of around 12,800 years. Because much of the Adirondack dome was ice-free by 13,000 years ago, mammoths or mastodons may well have roamed these uplands for several centuries before vanishing from history. Perhaps it's only a matter of time before someone finds a massive leg bone or tusk here.

What other animals lived in the Adirondacks back then? Some of the larger candidates are known from elsewhere in the Northeast at the time but, like the elephants, are now extinct. They included short-faced bears, the largest predatory mammals ever to live in North America. The biggest *Arctodus simus* were roughly twice the size of grizzlies, weighed nearly a ton, and could run almost as fast as horses. The omnivorous and adaptable monster-bears lived throughout the continent for at least a million years until roughly 12,800 years ago.

Another potential early Adirondacker was a form of giant beaver who measured seven feet long from snout to tail and gnawed trees down with six-inch teeth. They were not just hefty versions of today's beavers but another creature altogether in their own genus, Castoroides. They lived from Ontario to Florida for millions of years before disappearing around 12,000 years ago.

Many other kinds of large mammal vanished around the same time, as well, including peccaries, giant ground sloths, and stag-moose, but the cause remains unclear. Some scholars suggest that climate change at the end of the last ice age was responsible, but such a blanket statement is questionable.

The relatively sudden disappearance of mega-mammals across the continent cannot be easily attributed to a simple death-by-warming story. They survived millions of years of environmental changes, including multiple deglaciations that resembled the last great thaw. They successfully adapted to both warm and cold habitats, and when climatic zones retreated northward with the ice, the more southerly populations needed only to shift their geographic ranges accordingly. Furthermore, the kinds of plants upon which many of these animals depended did not disappear altogether but merely shifted their own home ranges along with the changing climatic zones. Smaller mammals who shared habitats with the largest ones survived, too, including species of deer, bears, and beavers we know today.

Climate change isn't likely to target supersize mammals more than most other species, but a discriminating hunter might. A recent study headed by Matthew Boulanger, an archaeologist at the University of Maryland, showed that people were living in the Northeast around 13,000 years ago and therefore overlapped with the final days of large mammals. If overhunting was indeed responsible for the extinction, then it happened on a massive scale in ways that have yet to be determined. Definitive kill sites that link hunting implements to butchered elephant bones and such are rare to nonexistent in the Northeast, but they have been found farther west, and a recent isotopic study of human remains from Montana shows that mammoths and other large mammals were major parts of the diet there around 13,000 years ago. Even so, "people did it" can also feel like a hard pill to swallow.

We may never know exactly why we need not carry mastodon bells or *Arctodus* spray in the Adirondacks today as hikers out west do with bear-safety gear in grizzly country. Nonetheless, there can be no doubt that climatic changes did help shape the Adirondack ecosystems and cultures that followed the final days of the great mammals, as we shall soon see.

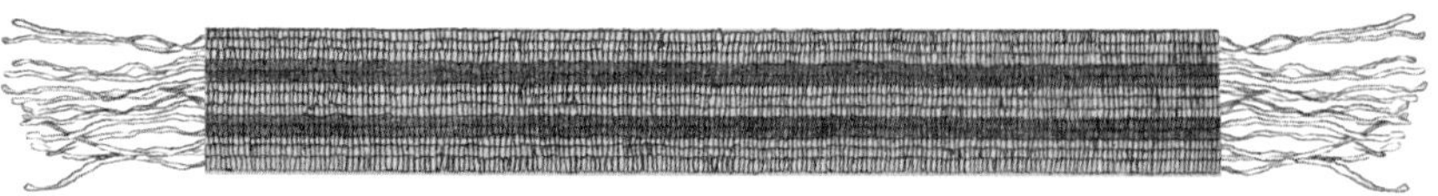

By 11,000 years ago, give or take, Adirondack climates had grown substantially warmer. In response, a more diverse array of birches, oaks, pines, and other kinds of tree began to replace the boreal parklands with mixed hardwood-conifer forests.

Many of the largest mammals were extinct by then, but plenty of others thrived in those forests. Their descendants are still with us today, although some were wiped out by people in recent centuries. Among the recently lost species were caribou, some of whom were still living in northern New England during the 1800s. They were probably common in the Adirondacks, too, as were mountain lions, timber wolves, and lynxes, who were lost before the end of the nineteenth century. Moose and what we now consider to be normal-sized beavers were also decimated but have since rebounded, and many other creatures—from turkeys to turtles, fish to foxes—lived in the region for millennia and left their remains in archaeological sites across the Northeast. All were potential prey for the first human residents of the uplands.

Meanwhile, other dramatic events were unfolding in the surrounding low country where changes in the land and lakes supplemented the diets of early Adirondackers with, of all things, seafood. When the heavy ice sheet receded the land rebounded, but only very slowly. Before the down-warped beds of the Saint Lawrence and Champlain Valleys could respond much, they were submerged by saltwater. An arm of the Atlantic Ocean penetrated far inland, forming a great tidal estuary around the northern rim of the Adirondack dome. From 13,000 years ago until about 9,000 years ago that estuary, now known as the Champlain Sea, covered more than 20,000 square miles and teemed with marine life.

In 1849, a railroad construction project unearthed the skeleton of a beluga whale in Champlain Sea deposits near Burlington, Vermont. In 2009, students in a SUNY Plattsburgh geology class discovered the bones of a ringed seal in 12,500-year-old marine sediments near their downtown campus. Other deposits have revealed the remains of walruses, narwhals, bowhead and finback and humpback whales, two more species of seal, and enormous brown bears who might have scavenged the beaches for whatever carcasses washed ashore. The shells of blue mussels and chalky white clams still litter gravel pits near Lake Champlain, all indistinguishable from shellfish who live along the North Atlantic coast today.

Dense, gray marine clays found beneath the cities of Ottawa, Montreal, and Quebec show that those areas were also submerged by seawater then. When the deposits become wet from heavy rainfall or groundwater seepage they can sag or slide, damaging roads and buildings in the process. Those same clays were more beneficial to inhabitants of the distant past because they are ideal material for making pottery. Ceramics made

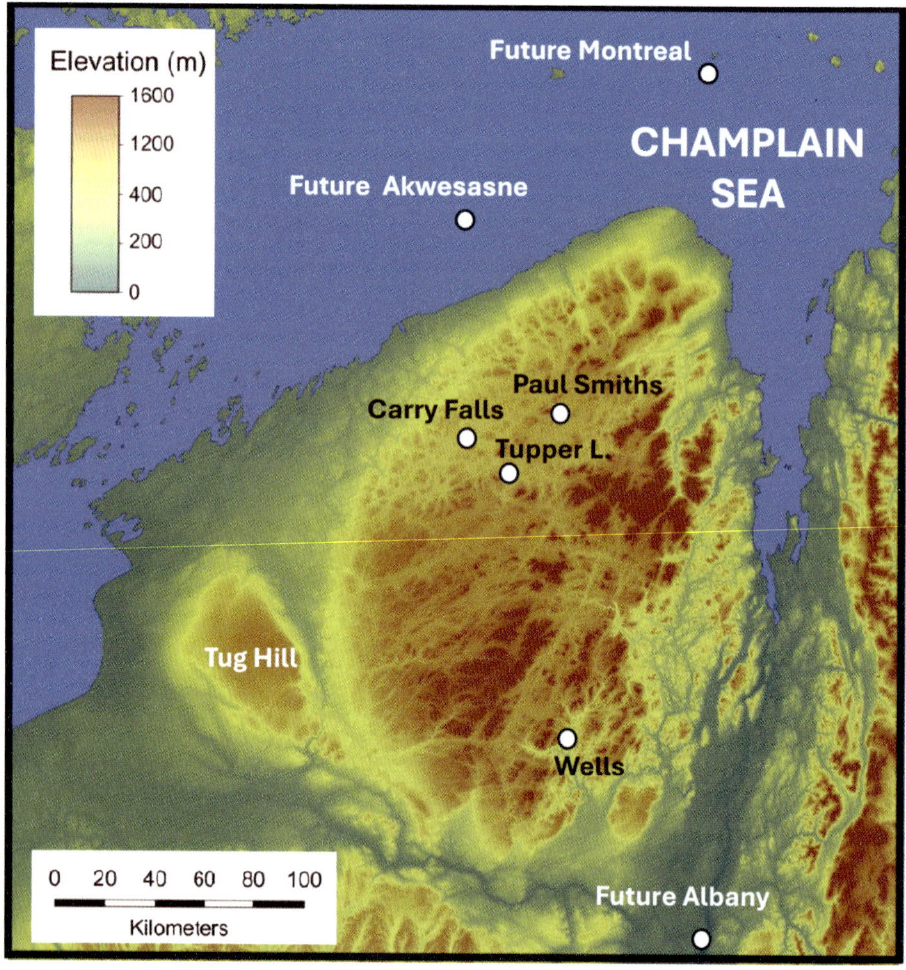

Above: Site map showing the Adirondack uplands and Champlain Sea roughly 10,000 years ago. Future locations of Montreal, Akwesasne, and Albany are labeled in white. Approximate locations of upland sites where items linked to Paleoindian and Early Archaic cultures have been reported are labeled in black. The base map and extent of the Champlain Sea shown here were provided by David Franzi.

Opposite: Paul Smith's College students using a Russian peat sampler to collect a core from Scarlet Bog. Seven sturdy meter-long rods were used to push the sharp-tipped sample barrel (foreground) down to the base of the peat deposit.

from Champlain Sea clay would link people of the uplands to the lowlands for thousands of years, as we shall see in later chapters.

After about 8,000 years ago, more hemlock groves appeared among the pine stands, and more deciduous trees such as ash, elm, beech, and maple further enriched the forest community. Climates in the Northeast were warmer than during the boreal parkland phase and were also becoming moister. Core samples from local wetlands show that the wetting trend was happening in the Adirondacks, too.

In the midst of a thick stand of spruce and fir near Paul Smith's College lies a beautiful little clearing that resembles a plush red carpet. The carpet is a "quaking bog," a spongy, waterlogged mat of sphagnum moss that fills a former glacial kettle hole in the sandy deposits below. Such wetlands are named for the bouncy trampoline effect they produce when people walk on them. The soggy upper layers are soft and buoyant, floating on a layer of water below. At first, the sensation can be disconcerting as the seemingly solid ground beneath your feet sags and ripples, but it is perfectly safe. Well, in theory anyway. When bringing anxious students there to sample, I've sometimes quipped, "If you do fall through, at least we'll always know where you are."

Some of the sphagnum mosses in this bog are lime green but most are bright red and so abundant that one awestruck student named it "Scarlet Bog." It's a captivating world in miniature to explore, brimming with tart cranberries, bug-ensnaring sundews, and insectivorous pitcher plants that Indigenous peoples of the Northeast variously called "turtle socks" or "frog shoes." But even more enthralling is the sense of history it conjures when we probe the depths of Scarlet Bog itself.

A clump of sphagnum plucked from the bog surface exposes the blanched remains of ancestors below, well preserved by the acidic chemistry of the bog water. Press a core sampler down through successive generations of moss and you can collect pieces of bygone worlds. In some places where such wetlands are drained or excavated, the bodies of mastodons or even humans may emerge, as thoroughly mummified as the ancient mosses they had lain within. At Scarlet Bog, we have pushed more than twenty feet of rod and pipe down into the fibrous peat mat until dense sand halted our sampler. Those peat cores have not brought up any mastodons or people yet, but they routinely yield seeds, conifer needles, glossy beetle wings, and other representatives of the distant past.

Peat cores from local bogs also provide glimpses of natural climatic shifts that could have affected early Adirondackers. The lowermost peat layers in Scarlet Bog are about 8,400 years old, and the basal sediments of most other wetlands we have sampled thus far range in age from about 8,500 to 5,500 years. Wetlands require wet conditions, of course, so the sequential formation of these moist habitats one after another means that groundwater levels have progressively risen as climates have become wetter over the millennia.

Despite the long-term wetting trend, the list of common forest tree species in the Adirondacks has remained relatively stable with the exception of one mysterious event that complicates the story. About 5,000 years ago, something terrible happened to the hemlocks. Their pollen nearly disappeared from Adirondack lake sediments and peat deposits for centuries, only approaching former abundances after several millennia of recovery. Much the same thing happened throughout the Northeast, but the cause of the catastrophic hemlock decline remains a mystery.

Glimpses of what happened then come from peat cores that were collected from Bloomingdale Bog in 2005 by Katharine LeBouef, a graduate student at Lehigh University. Hemlock represented more than a third of the pollen assemblage from 7,000 years ago until about 5,000 years ago, after which it crashed to nearly zero. Droughts and fires could have prolonged the decline once it started, but hemlock-eating insects or a disease were more likely to have been the initial trigger. As LeBouef noted in her dissertation, the mysterious hemlock disaster "is one of the most well-studied vegetation changes of the (last 10,000 years), and yet we still don't fully understand it." We can—and will, in a later chapter—only imagine how Indigenous Adirondackers experienced that event in real time.

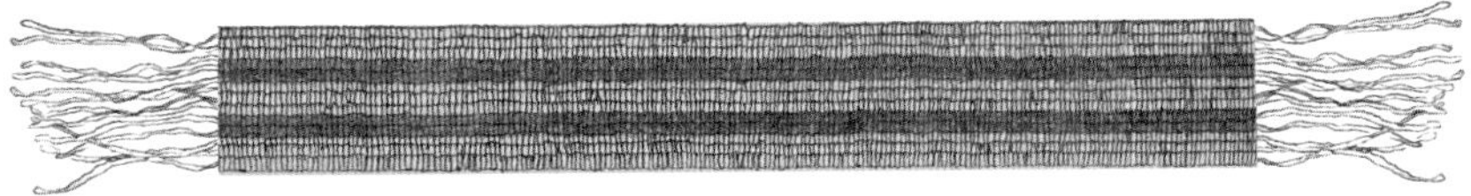

We now have a 13,000-year overview of the environmental settings in which hundreds of generations of Adirondackers lived out their lives. The rest of this book will explore that history in more detail.

As the tale unfolds, the lifestyles of the people we meet along the way can be linked to four main cultural periods that have been defined by archaeologists. The sequence of those periods is illustrated on the chart shown here along with very rough approximations of their timing.

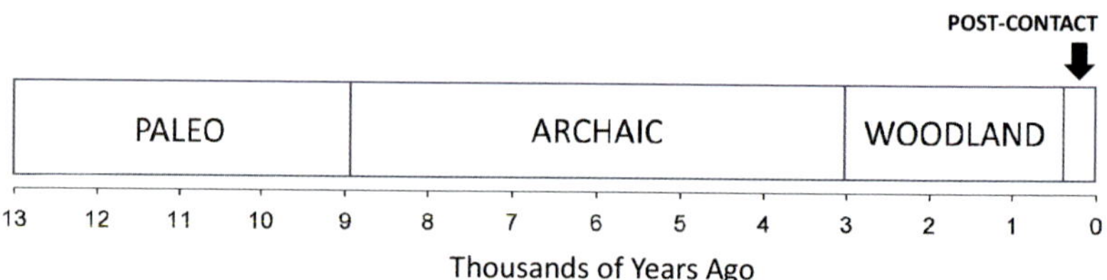

Combining those cultural periods with local environmental history, we can set a basic framework of four stages for the chapters to follow.

Stage 1: Approximately 13,000 years ago, the postglacial climates were much cooler than today. Boreal parklands with open herb-rich tundra, spruce thickets, and scattered pines covered the newly exposed landscape, and the Champlain Sea submerged much of the adjacent Saint Lawrence and Champlain Valleys. Musk oxen lived in the Adirondacks, and large mammals who have since become extinct, such as mastodons, mammoths, giant beavers, and short-faced bears, were present in the region if not in the uplands themselves. Caribou, brown bears, and many other mammal species were probably also present. The hunter-gatherers who arrived sometime between 13,000 and 12,000 years ago are referred to by archaeologists as "Paleoindian" peoples.

Stage 2: Between about 11,000 and 8,000 years ago, climates became warmer and mixed hardwood-conifer forests gradually replaced the boreal parklands. Pines and an assortment of deciduous trees dominated the hardwood-conifer forests, and a wealth of mammals including black bears, deer, moose, caribou, and beavers probably thrived in the uplands. The Champlain Sea continued to support seals, whales, fish, and shellfish until uplifting hoisted Lake Champlain above sea level as a freshwater body about 9,000 years ago. Hunter-gatherer lifestyles transitioned from Paleoindian to so-called "Archaic" cultures sometime between 10,000 and 8,000 years ago.

Stage 3: After about 8,000 years ago, Adirondack climates grew slightly cooler and significantly wetter, with water levels rising in the lakes and new wetlands forming in low areas. Various deciduous trees and eastern hemlocks became common in the diverse forest community, but the hemlocks declined suddenly about 5,000 years ago, which might have helped make some species such as beech and maple more numerous. Pottery and horticulture became widely used during the "Woodland" cultural period that began about 3,000 years ago. However, local residents continued to hunt, fish, trap, and gather wild foods in the uplands as their ancestors had done through the ages.

Stage 4: The arrival of non-Indigenous people during the "Post-Contact" period of the last five centuries caused massive environmental disruptions ranging from species extinctions to land clearance. Those disruptions were also accompanied by extreme cultural upheaval, the echoes of which still reverberate among us today.

We are now ready to begin our long journey through the human story of the uplands. Our first stop will be a pleasant summer day on the knoll overlooking Lower Saint Regis Lake some 12,000 years—or 240 lifetimes—ago.

OLDER THAN THE TREES
(12,000 Years Ago)

On this bright summer morning 12,000 years before our own time, the knoll overlooking the lake is devoid of trees. Instead of forest, it wears a patchwork cloak of velvet-green mosses, fluffy cotton grass, and clumps of ivory-white and golden alpine flowers. Native bees and hoverflies buzz among the ground-hugging blossoms, drawn by the sweet scents that dawn's warmth has lifted from them. The gray bluffs and bald pate of what will later be called Saint Regis Mountain press starkly against bright blue sky on the western horizon, and a textured quilt of soft tundra herbs, bristling spruce thickets, and a few lone pines decorates the lower slopes.

We are not the only people here, though. Voices, laughter, and soft footfalls are approaching from the mainland.

Soon a family of four joins us on the knoll, unaware of our virtual presence. They are two young parents with a girl and boy, both of whom appear to be about ten years in age. All are clad in smoke-scented caribou skins. Their feet are wrapped in moose-hide moccasins and their long black hair dances in the wind.

Father wears a necklace of enormous bear claws. He sits down beside Son on the sandy rim of the knoll where a brisk breeze keeps mosquitoes at bay. Behind them, Mother and Daughter stoop for sweet, apricot-colored cloudberries. And farther back on the lee side of the knoll, conical shelters covered with caribou hides mark an encampment that the family shares with relatives. It is one of several high, dry places that this band favors while roaming a vast territory in search of food. Their home range extends all the way down to the edge of the Champlain Sea, where seals and tasty shellfish abound and where people gather on occasion for trade, advice, and companionship. That is where Father won the claws from a fierce brown bear with the aid of other hunters. The great beast had made the mistake of trying to claim the carcass of a moose whom the men had just slain in a marshy meadow where the outlet river from this same lake meets the sea.

The boy is working on a piece of gray Hudson Valley chert with an antler tip and a leather pad to protect his palm. He gingerly presses the bony tip against the edge of the shard with the hesitancy of a beginner, but nothing happens. He tests another spot, then another, until a thin flake finally breaks loose with a sharp click. Father instructs him between tries in a language that we don't understand, but his body language is clear. He is growing impatient.

Father leans over, takes hold of the flaking kit minus the pad, and begins to work the grooved shard himself. Click. Click. Click. He tips his hand repeatedly to let tiny flakes of chert drop away, but not for long.

Click. Click. SNAP!

Son giggles, then checks himself when Father aims a testy glance at him. The practice shard, inexpertly shaped in a way that made proper finishing difficult, has broken in two and drawn a bit of blood, as well.

Disgusted, Father tosses the ruined dart point aside, rises, and walks down to the water's edge to wash the cut on his palm. Son follows, hiding a smile.

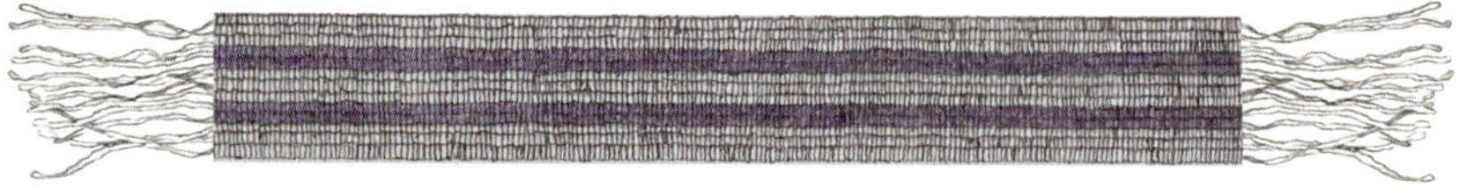

Twelve thousand summers later, my wife Kary and I visited the knoll to savor a bright sunny day, the scenery, and the rich aroma of red pines who have long since replaced most of the tundra herbs. We had recently begun to explore the deep human history of the Adirondacks with help from John Kahionhes Fadden, then director of the Six Nations Indian Museum, and Tim Messner, an archaeologist at SUNY Potsdam.

As Kary wandered the pine grove behind me, I sat on the crest of the knoll, taking in the view of the lake and mountain and imagining the people who had cherished this place long before we arrived.

Lost in my musings, I didn't hear Kary calling to me at first. Then she called again.

"I think I've found something."

As she walked closer, I saw a small triangle of gray chert resting on her open palm. Messner would later identify it as a projectile point of a sort that could be anywhere from centuries to millennia in age.

We were astonished to find and touch such a tangible link to Adirondackers of the distant past right here on "our" knoll. We spent the next hour on our hands and knees, gently sweeping aside fallen conifer needles, empty pistachio shells, and deceptive flakes of pine bark. I soon uncovered a darker gray triangle of chert amid the debris. It was larger, and the base appeared to have been broken off. More sweeps turned up a rectangular piece with the same color and a matching edge.

When we joined the two pieces together, they formed a slim blade that was as long as my thumb. It had been roughly flaked on both faces and thinned at the base with a short groove running up the middle. Similar grooves, also known as "flutes," are found on projectile points that were made throughout much of North America roughly 13,500 to 11,000 years ago.

Jonathan Lothrop, an expert at the New York State Museum, would later conclude that the blade seemed to be an unfinished dart or spear point. During the deglacial period, bows and arrows were not yet widely used in the Northeast. Instead, piercing weapons were either thrust by hand or hurled more forcefully with the aid of a throwing stick. The stick, later to be known by the Aztec name "atlatl," served as an extension of the throwing arm and was sometimes adorned with a stone weight or "bannerstone" that gave it extra heft and balance. Atlatls have been used by people throughout the Americas, Eurasia, and Australia for many thousands of years, representing a nearly universal technological heritage.

If Lothrop's surmise was correct, then the broken point would be one of the oldest known relics of human presence in the Adirondack uplands. However, its incomplete workmanship made it difficult to identify with certainty. Nor could we know when or why the break happened. Similar breaks are common among unfinished fluted points, which is not surprising considering their thin, delicate shapes and the skill required to produce them. This break could have been a reason to discard a point in progress long ago, but it could also have been made recently under the foot of a careless tourist or student. Despite this uncertainty, there is no such doubt about the identity of another projectile point that was found near Tupper Lake, less than an hour's drive to the west.

In 2007 a hydroelectric dam on that lake's Raquette River outlet needed repair, so water levels were lowered by several feet to facilitate the maintenance. As a result, the New York State Historic Preservation Office requested that scientists from the Public Archaeology Facility at Binghamton University examine the shoreline while the lake remained low.

Daniel Seib, the leader of the team, recently described the experience for me. "The dam had raised and lowered the water levels of the reservoir over and over for more than a century, so waves had eroded the shoreline," he said.

Their task was to search the beaches for ancient items that had been exposed along the water's edge. They weren't disappointed. Over the course of a month they found a dispersed trove of stone implements representing thousands of years of human presence at 1,500 feet elevation.

Perhaps most exciting of all was an object that was found in addition to assorted flakes of chert and quartzite, sharp-edged scrapers that may have been used to prepare hides, slender stone awls, and signs of former campfires. Lying amid the rocks and mud was a tapered chert point four inches long with a shallow groove running up from the base on both sides of it. Dan immediately recognized the fluting as a diagnostic feature of some of the oldest known hunting tools in North America.

Although radiocarbon dating can work well on materials that are rich in carbon such as wood or bone, it doesn't work on stone points. Their ages must instead be inferred through their association with more readily datable organic materials in archaeological excavations. That is just one of many reasons why amateur collectors who carelessly dig through sequential layers of sediment for plunder risk destroying history. However, distinct styles of projectile point have now been dated in other regions of the

Northeast, so identifying the specific style can also help determine its age. Jon Lothrop would later conclude that the Tupper Lake point was of the "Barnes" or "Michaud-Neponset" style that was commonly used in the Northeast between 12,200 and 11,600 years ago.

The importance of finding a fluted point along with other tools whose ages spanned the last 12,000 years was not lost on the survey team, as their official report made clear.

"Previous finds suggest major waterways were used by Paleo groups to penetrate the interior of the state. Despite this, it has been suggested that the Adirondack region was not utilized by these groups. . . . The presence of Paleoindian lithics in such close proximity to a Middle Woodland feature indicates that not only were Native American groups in this valley, but they returned repeatedly."

Seib's team was able to examine only a small fraction of Tupper Lake's shoreline, and he believes that a complete survey could have found many more signs of human presence from long ago. "Back then it was a perfect place for finding food," he said. Before the dam flooded the site there were wetlands, tributary creeks, shallow oxbows, and deep channels in which all sorts of animals and plants would have thrived. The new finds made it clear that the food-rich setting had supported generation after generation of hunters, fishers, and gatherers at this upland site.

How does Daniel Seib, as an archaeologist, feel when he thinks of the undiscovered legacies of human history lying beyond reach in the depths of the reservoir? He never got to see what lay farther offshore, where, he believes, most of the remaining relics are not only submerged but also buried under thickening blankets of lake sediment.

"I'm all for it," he said. "The water and sediment protect everything from amateur collectors like the ones we saw taking things from the beaches we were working on." He and his teammates were visible to boaters and other curious passersby, and upon returning to one site that they had previously flagged for study they found that someone had removed the marked objects.

In another instance, a kayaker approached Seib to show him a rectangular stone with winglike flanges on either side and a hole bored through the center of it. She claimed to have picked it up elsewhere on the lake, but Seib thought it was more likely that she found it among their study sites. The odd-looking artifact was a bannerstone that had once been fitted to an atlatl, and of a type that had not yet been formally documented in the Adirondack uplands. Nor will that one ever be.

The loss of the pilfered bannerstone was a missed opportunity to learn more about the history of hunting technology in the North Country. Atlatls are believed by archaeologists to have been brought to North America long ago by ancestors who arrived from Asia. However, the remains of those hunting tools are rare because wooden implements can decay quickly. Durable bannerstones are the oldest known remnants of atlatls from the Northeast, and several that were unearthed from a site in southern Ontario are thought to be about 10,000 years old. Could the kayaker who took the Adirondack bannerstone have made off with some of the earliest evidence of weighted atlatl use in the uplands?

"For me," Seib explained, "it's enough to know that the submerged artifacts are still out there, safe and sound, even if we never get to study all of them." Although the number of items found at Tupper Lake was limited by the short time frame available for the shoreline survey, Seib's work provides more than enough evidence to refute claims that Indigenous people were absent from the uplands. Moreover, additional information refutes the claim, too.

According to Lothrop, at least two other fluted projectile points have been identified thus far in the uplands. One was found near Carry Falls Reservoir several miles downstream on the Raquette River at an elevation of nearly 1,400 feet. Although it was too fragmented for positive identification, it is believed to be 12,800 to 12,200 years old. The other point was found by a private collector near Lake Algonquin in the southeastern Adirondacks. Like the one from Tupper Lake, it was accompanied by additional stone implements whose ages ranged from the deglacial period to the last few centuries. Such finds confirm that people were present in the Adirondacks very soon after the ice sheet uncovered the uplands roughly 13,000 years ago.

These silent heirlooms provide powerful evidence of widespread human presence among the stark boreal parklands of the Adirondack dome that, in its earliest days, lacked the more protective, resource-rich forest cover we know today. For me, it brings to mind a song popularized by the late John Denver in which the lyrics say, "life is old there, older than the trees." Here in the Adirondacks, those words are literally true.

How might a fluted projectile point have been used long ago in the Adirondacks? While we tend to focus on the drama of large animal hunting when we envision those times, it is important to remember that First Peoples consumed a wide range of other foods, from berries and tubers in summer to ptarmigans and snowshoe hares in winter. However, hunting of large prey was indeed a major source of food in the deglacial Northeast, as it has been for long ages among peoples of the Arctic. Therefore, it's worth considering some of the ways in which spears and darts could have been used to sustain the earliest residents of the Adirondacks.

There is little unambiguous evidence for the pursuit of mammoths and mastodons in the Northeast, but we can be more certain of caribou hunting, which was commonly practiced across northern North America long ago. Some of the most spectacular insights into how such activities were carried out come from the high mountains of the Yukon Territory, where ancient snowfields are thawing under the onslaught of global warming. As they melt, they disgorge signs of caribou hunting of the sort that could have occurred in the Adirondacks, as well.

In 1997, archaeologist Gerald Kuzyk scrambled up to a melting snowfield at 6,000 feet elevation on a peak in the Yukon and found a slender piece of wood sticking out of a brown-stained patch of ice. It was a 5,000-year-old fragment of an atlatl embedded in an ancient deposit of frozen caribou dung. Since then, members of the Yukon Ice Patch Project

Above left: The first projectile point found on the lakeside knoll by Kary Johnson. *Above right:* A possible unfinished fluted point found in two pieces on the knoll. *Lower right:* The crest of the knoll where the points were found.

Opposite: The fluted projectile point found by Daniel Seib's team at Tupper Lake.

Below left: Students using atlatls for target practice. The fletched butt-end of the dart bears a shallow depression into which the pointed tip of the atlatl fits. *Below right:* When the dart is thrown, the atlatl essentially lengthens the thrower's arm and thereby applies greater force to the dart.

have studied many such sites, with both Indigenous and non-Indigenous investigators participating. They have found hundreds of beautifully preserved hunting tools ranging in age from 9,000 years to modern, along with layer upon stinky layer of caribou poo.

The frozen deposits protected the implements so well that wood, fletching, and sinew bindings that would normally have decayed away are still intact. The association with droppings is also consistent with testimonials from keepers of traditional knowledge and recent scientific accounts in which female caribou travel high up into the mountains in summer to find snowfields to bed down on with their calves where the chill discourages insect pests. That predictable behavior offered prime opportunities for early hunters.

Such information from the Yukon ice patches can help flesh out a reasonable account of early hunting in the Adirondacks, as we shall soon see. But another remarkable find from the Great Lakes reveals another way in which early hunters captured wary, fleet-footed caribou.

In 2014, underwater archaeologists from the University of Michigan were using remote sensing technology to study formerly dry terrain beneath Lake Huron when their equipment detected a long line of boulders on a submerged ridge. They had found a 9,000-year-old hunting fence, an artificial barrier of the sort that Indigenous peoples of the North used to guide fleeing animals such as caribou into a confined space where hunters lay in wait. Closer inspection by divers and remote-operated vehicles revealed stone projectile points and several rock structures near the fence that could have been used as blinds.

With these studies in mind, we can now better imagine how a long-ago caribou hunt might have proceeded on the slopes of Saint Regis Mountain. In this imaginary scene, the residents of the knoll are about to use a hunting fence of the sort that would later be discovered under Lake Huron. Let the action begin.

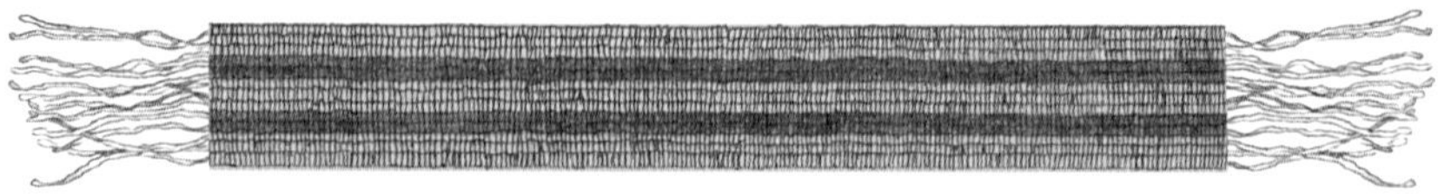

It is still early summer, and the people whom we met on the knoll have joined the rest of their local band on a trek to a bowl-shaped basin on the northern flank of the mountain. After crossing four miles of sparsely wooded terrain, the party of two dozen approaches the place where a herd of caribou were spotted as they climbed uphill to a bed of remnant snow that has been shaded by the basin walls. These experienced Adirondackers know just how to seize the opportunity.

Fording a stream at the base of the mountain, the hunters push uphill through dense thickets of shrubby spruce, moving with purpose and speaking only in whispers. The men carry slender atlatl darts whose shafts they had fletched with raven feathers and tipped with sharp chert points on detachable foreshafts. Two of the women carry a long roll of netting woven from sturdy threads that were extracted from dogbane stems. The children control the group's four terrier-sized dogs, keeping them leashed and muzzled for the time being.

Before leaving the protective cover of the the thickets, everyone pauses to plan the hunt. This will be a team effort, and each member of the group will play an important role. The women and kids will continue uphill on the outer flank of the basin with the dogs, well hidden from the prey on the other side. They will then conceal themselves along the ridgeline above the snowfield where the caribou are resting with their calves.

Meanwhile, the men will creep up to the thinly vegetated mouth of the basin where two long lines of boulders nearly meet in the middle. The ages-old barriers have lengthened and become increasingly solid as successive generations of hunters placed more rocks on them, and additional fencing of brush and branches was added recently. A twenty-foot gap between the downward-angled walls allows a stream to trickle down where the caribou recently climbed up. On either side of that gap lie heaps of stones behind which hunters can hide with weapons and netting at the ready.

High above the men, the herd rests quietly on the dung-darkened snow, unaware of the danger. The wind is coming from the west, carrying the telltale odors of humans and dogs away from the caribou. The anticipation is palpable.

Suddenly, loud mayhem erupts at the top of the basin. A dozen women and youngsters charge downhill, shrieking and shouting alongside their howling dogs. A hundred caribou lunge to their feet and flee in what seems to be the safest direction: downward. In a panic, they race blindly toward the fence below. Those on the outer margins veer toward the center as the walls funnel them inward, adding to the jostling confusion in the middle.

When the torrent of caribou bursts through the opening in the walls, the trap is sprung. Several men rise from behind their blinds and hurl darts into the stampeding herd. Father is particularly skilled with the atlatl, and his movements are fluid and precise. He swings his right arm back with the foot-long stick gripped tightly in his hand and a six-foot dart pinned to it by his thumb and forefinger. A depression in the base of the dart rests snugly against an antler tip mounted on the far end of the atlatl. He swings his arm forward, releases the dart, and feels the atlatl tip push the projectile forward with devastating force. A sharp bleat, and another caribou falls.

Within seconds, most of the animals are past the gap and beginning to escape. Two teenaged boys rise and hoist the net they had placed on the ground just below the opening. Two caribou calves are quickly entangled in it, and the boys rush at them with spears in hand.

Now, the more tedious work begins. Several escapees have been wounded and need to be tracked down. Loose shafts of darts that struck their marks and left their foreshafts embedded in animals who fled are also retrieved. The carcasses of the fallen will be butchered and taken back to camp on the knoll, where firewood is more plentiful. There the meat will be sliced into thin strips and dried on wooden racks over smoky fires to preserve it for future use.

Women and girls will stake the hides on the ground, scrape off the excess fat and tissue with tools of stone or bone, and tan the skins with caribou brain paste. They will press, tug, and stretch the hides until they are soft and supple enough for garments, bedding, and coverings for dwellings. Men and boys will sharpen their projectile points and make

new darts to replace those that were lost or broken. Caribou tendons will be turned into sinew threads; bones into awls, needles, and fishhooks; and antlers into flaking tools and atlatl tips.

The hunt has been a happy and important event for these residents of the knoll, but not one that occurs every day. Caribou can learn, too, and using the basin as a trap too often could prevent the herds from returning in the future. Moreover, there are plenty of fish to catch, hares and ptarmigans to snare, herbs and berries to pick, and rich marine habitats of the Champlain Sea to exploit downstream.

After generations on this territory, the people know it well. They know how to navigate it over great distances, where and when to find the plants and animals who sustain them, and how to avoid overusing the ecosystem. As long as the group works together as a unit, whether bound together by blood or by agreement, they will have a long, sustainable future in their expansive mountain home.

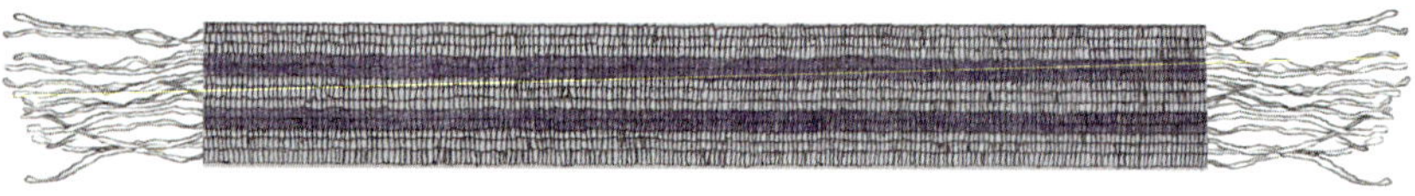

How much of the region the first Adirondackers considered to be their home territory is a question that may never be fully answered through archaeological finds alone. The scarcity of sites dating back to the deglacial period and the light impact early residents had on the land have left little for us to work with. However, the presence of Paleoindian peoples in upland settings is well documented farther to the east. Ancient sites have been found in the White Mountains of New Hampshire and the Green Mountains of Vermont, and one was discovered at an elevation of more than 2,000 feet in the Taconic Mountains of southeastern New York. With fluted projectile points now documented in the Adirondacks, as well, we can be certain that Indigenous people lived in the uplands relatively soon after the land emerged from its ice cover. As we shall see in later chapters, they also continued to live here up to and including the present day.

People were present in the surrounding lowlands during those early days, too. One such site lies on private property on a floodplain just a few miles upstream from where the Ausable River flows into Lake Champlain. Over the millennia, floodwaters have deposited thick layers of fertile sediment that now support a productive local farm, and stone implements often turn up in the fields after the spring plowing. Herein lies a problem.

If this farm were located in Europe, such implements would likely be considered part of the historical heritage of the entire country and therefore owned by the nation as a whole. For instance, when a trove of Iron Age weapons was found in a privately owned wetland in Denmark some years ago, professional archaeologists were called in to protect and excavate them. They carefully mapped and excavated the deposit, published their findings widely, and archived the items in a museum where everyone could enjoy them, learn from them, and weave their stories into a proud sense of identity.

Not so in the United States if the discovery occurs on private land. The items collected near the Ausable River came from fields that had already been disturbed by decades of plowing, so there may have been little harm done in simply picking them up from the surface. But as a personal collection they have since been kept from casual public view. If they were to be sold off individually, inadvertently discarded, or mixed up with items from elsewhere, yet another page of history would be lost forever. Nonetheless, it is all perfectly legal.

Fortunately, many people who find ancient items recognize the importance of alerting reputable experts who can help protect and properly interpret them. For example, some of the earliest evidence of ritual behavior in North America was reported to archaeologists within hours of its discovery in southern Quebec. Brian Deller and Christopher Ellis, experts from Ontario, soon determined that the dozens of stone implements they unearthed in the owner's farm field had been shattered intentionally, possibly as a form of ritual "killing" or sacrifice that took place more than 10,000 years ago. In their summary published in 2001, the authors also noted that much of the excavation work was done by local volunteers and summer students.

Similarly, the Meadowcroft rock shelter site in western Pennsylvania, possibly more than 16,000 years old, was discovered by a local resident who informed experts about it before it could be looted or otherwise damaged. Jon Lothrop and several coauthors acknowledged such contributions of amateurs to the field of archaeology in the closing paragraph of a paper they published in 2016. They wrote, "Insights into earliest peoples of the Northeast depend in part on our collaborative efforts with avocational archaeologists," and added that "the extent to which we can successfully partner with the avocational community will determine in large measure future accomplishments in . . . the Northeast and elsewhere."

A member of the family who owns the Ausable farm site has shown their collection to an archaeologist from the University of Vermont, but with no formal study of the property available we can make only basic inferences about it. The site was occupied repeatedly over thousands of years, as evidenced by the large number of stone objects of varied ages. It might have been a popular meeting place for people who used the valley as a travel route to the uplands, or a seasonal source of fish who migrated up the river from the Champlain Sea. Perhaps the broad, flat plain was a convenient location for family gatherings, for sharing news and knowledge, or for trade.

Many other questions about the earliest Adirondackers remain to be fully answered, too. What was daily life like for them, for example? Archaeological evidence from other regions of the continent suggests that most people of those times shared their home territories with small numbers of friends or relatives because the land could not sustain large groups in any given location for very long. They most likely lived in portable lodges covered with animal hides that were easy to move from place to place. Their annual travels, which were likely made over hundreds of square miles, were not mere random wanderings, however. They used well-known routes among familiar locations that offered food, shelter, and other resources at particular times of year. As a result, the entire region became their home.

We can only speculate about the ways in which the seasonal rhythms of the land and waters governed the lives of the earliest Adirondackers. The Champlain Sea might have provided shellfish most conveniently in summer, along with the chance to surprise a seal on shore or scavenge a freshly beached whale from time to time. Berries, greens, and waterfowl could have been had at any elevation during the warmer months. Perhaps winter was a good time to trap furbearers, fish through the ice, and pursue moose or caribou among deep snows in the uplands. Whatever the seasonality of their activities was, it reflected a lifestyle that was closely attuned to the ecosystems and landforms of the Adirondacks.

Did these people also use watercraft to cross the Champlain Sea or pursue fish and seals there? Some of the stone gouges, axes, and drills they made suggest a capacity for woodworking, so it is possible that they constructed dugout canoes, wooden rafts, or hide-covered frame boats. But we will have to be fortunate indeed to obtain direct evidence of that because perishable wood, leather, and sinew rarely survive such long journeys through the ages.

What languages did the first Adirondackers speak, and what were their spiritual traditions? What songs did they sing around the fire at night, and what stories did they tell their children and grandchildren during the long, cold nights of winter? The Haudenosaunee account of the Creation resembles accounts that are treasured in Siberia, as well, suggesting that they might have been brought to this continent by people who crossed over from Asia in the distant past. If so, then early Adirondackers probably knew about Turtle Island, too.

Despite the vast reaches of unrecorded time that separate us from those long-ago people, we do know one thing for sure. Life is old here, older than the forests. And the many generations who lived in these uplands long after the last fluted projectile points were produced also made this landscape their home in unique ways of their own. We will explore their stories in the chapters to come, but first we must consider an intervening period that remains more heavily cloaked in mystery.

To do so, we will have to rely even more on reasonable inference and imagination because the First Peoples who lived in the Northeast after the Paleoindian period ended left fewer signs of their presence than usual. In fact, we know more about the land, lakes, and nonhuman species in those days than about the people who lived among them.

We'll begin by reconstructing the environmental stage upon which those stories unfolded, and then infer what might have transpired in that setting 8,000 years ago, viewing them as best we can through the misty veils of time.

THE MISTS OF TIME (8,000 Years Ago)

We now move forward in time to a summer morning 8,000 years before our own century. Pine trees and blueberry bushes have replaced most of the tundra flowers and cloudberries who once grew here on the knoll. The flanks of the mountain on the western horizon are also more thickly forested than they were when we last visited, but it is difficult to see them now from this vantage point. The air is clouded by a pungent gray haze, and the sun is a faint orange ball suspended behind a thick screen of smoke. The mountain is on fire.

This isn't the first lightning-struck forest fire of the year. Local climates are warmer than before; the forests are full of resinous, flammable pines and other coniferous trees; and this summer has been exceptionally parched. Tiny flecks of charcoal that filter down from the smoke today are joining others on the lake bed like stacked pages in an archive of Adirondack fire history that scientists will decipher thousands of years later. But that is of no concern to the residents of the knoll. The acrid smoke has become intolerable, and they have moved elsewhere to escape it.

Although we see nobody on the knoll at the moment, it nonetheless bears signs of human presence. Four circles of bare ground lie beneath the trees where bark-covered lodges recently stood. Fish scales and detached fins litter the water's edge. On the crest of the knoll, small flakes of chert tell of people who have shaped or sharpened their stone implements where the breeze keeps mosquitoes at bay and the view, when not obscured by smoke, is grand.

Most of these traces of human presence will soon vanish or disperse beyond easy recognition, as has happened with most of the possessions of Adirondackers from long ago. But for reasons that are difficult for us to discern across the intervening ages, even fewer personal items from this chapter of history will survive the journey to what we observers call the present day. Therefore, we will pay special attention to potential time travelers of that sort as we encounter them during our imaginary visit.

Where have the occupants of the knoll gone to escape the smoke? They've most likely moved northwest by following the lake's outlet on a downhill journey to a much larger river in the lowlands. In our minds' eyes, we locate a faint footpath beside the outlet stream, which will later be known as the Saint Regis, and follow it.

Alders, larches, and wet meadows that we pass along the river's edge seem normal enough, but to a trained eye the surrounding forests look slightly different from those of our own time. They are richer in pines and oaks, who thrive under relatively dry conditions. Here and there, a sunlit clearing invites travelers to stop and pluck juicy red raspberries. Equally inviting to the people who passed ahead of us are herbs who offer leafy greens or edible seeds as well as tangles of dogbane whose stems provide fibers that can be used to make sturdy cordage.

The travelers whom we seek would have had little trouble finding animal prey along this route to and from the lowlands, too. The cloven hoofprints and droppings of deer and moose are everywhere along the river corridor. About twenty miles from the lake as the raven flies, the swelling river spills over a cliff at a spot that will later be called Saint Regis Falls. Plunge pools at the base of the falls conceal brook trout who can be readily hooked, netted, or scooped up in baskets. However, thus far we have found nothing but the footpath itself to reveal the low-impact human presence here.

We have now traveled a meandering route that would have taken a band of unhurried Adirondackers several days to traverse on foot while hunting, fishing, and foraging. The land slopes gently downward beyond the falls to where the rim of the Champlain Sea once lay. The melting of the great ice sheet has let the land slowly rebound several hundred feet upward, lifting Lake Champlain and the Saint Lawrence River above sea level as their formerly salty waters progressively freshened. An occasional outcrop of gray marine clay, sprinklings of old clam shells, and the odd seal bone whisper faint memories of those earlier days, but the sandy coastal beaches upon which earlier Adirondackers once camped are now cloaked in forest.

As for the people who call this watershed home, where are they now? A thin plume of campfire smoke ahead offers a clue. Where the outlet river meets the Saint Lawrence, we find a meadow containing four circles of packed soil and a hearth that has ignited a small ground fire that still smolders beside it. The inhabitants of our knoll have been here recently.

They probably know this place well. Several months ago, residents of the region and travelers from afar gathered here to catch fish, possibly shad or alewives who migrate up rivers such as this in spring. That kind of seasonal bounty has long encouraged widely dispersed people to converge at river mouths or waterfalls and feast together while the food lasts. In doing so here, they could also share news of hunting conditions, trade food or tools for chert from distant quarries or bits of copper from the shores of Lake Superior, visit relatives, and find new friends, lovers, or spouses.

Thousands of years later, the mosaic of rivers, marshlands, and islands near this location will support a sizable community that Haudenosaunee residents call Akwesasne (roughly, "Where the Partridge Drums"), perhaps in reference to local wildlife or to the drumming sound of rapids there. During our imaginary visit, however, the scene before us is currently empty of humankind except for the recently vacated floors of the lodges. If the residents of our knoll are no longer here, then the next best option for locating them is to follow an alternative route back to the uplands. That is where we find them at last.

A nearby river, later known as the Raquette, drains a watershed adjacent to that of the Saint Regis, and it can easily be reached from here by a short overland traverse. The families we seek have been following the Raquette upstream to what will later be known as Tupper Lake. There they have erected their bark-covered lodges on a pine-clad hillock overlooking a marsh where fish, ducks, and turtles are abundant and where deer and moose often leave the forest to drink and graze.

The families are using this spot with the permission of relatives who normally occupy the area. Everyone understands the danger and inconvenience of forest fires, and nobody has begrudged these people a temporary refuge while they wait for the next heavy rainstorm to quench the flames near their preferred lake. Out of courtesy for their hosts, whose livelihoods depend upon what this section of watershed provides, their stay will nonetheless be brief.

The sun is setting, and we have little time left before our brief visit ends. How fitting to have spent most of this smoke-filled day seeking these people in much the same way that archaeologists of our own time will seek them across a clouded gulf of history. And as luck would have it, we are about to experience one of the rare moments in which a durable relic crosses that vast temporal gap.

A teenaged girl dressed in buckskin has passed us on her way to the nearest lodge. Unaware of our presence she pauses, stoops, and picks something up from the ground before entering.

Moments later, the girl reappears. She hurries back to where she had paused and replaces the object exactly where she had found it. From the doorway, an elderly woman who appears to be her grandmother calls out to her as if explaining something. The girl rejoins her, a broad sheet of bark slides into place over the door, and we are alone again in the dusky forest.

Moving closer, we discover the object of the girl's interest: a pyramid-shaped dart point made of dark gray chert that is nearly invisible among last year's pine needles and cones. It is about half the length of its maker's thumb and appears to have been abandoned because one of the notched corners at the base is broken, making the point useless for hunting. Perhaps the girl considered taking it as a keepsake, but her grandmother apparently reminded her that they are guests here and should not remove anything that doesn't belong to them.

As usual, Grandmother is right. Although the broken point can no longer make atlatl-driven flights at deer or moose, its sharp edges can still scale a fish, slice leather, or scrape a fresh hide. From our perspective, however, its potential value as a traveler through the ages outweighs that of any short-term use to its maker. With luck, it could find its way into the hands of someone who will recognize it thousands of years later and thereby bear witness to the Adirondackers who lived in this area long before our time.

And that is precisely what happens.

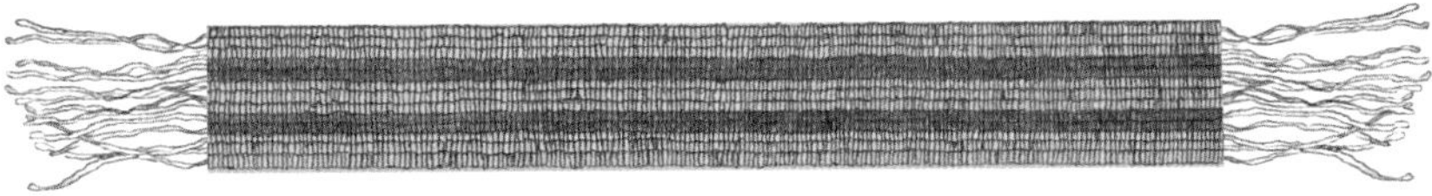

On a freezing cold, blustery morning in March, Kary and I arrived at the Public Archaeology Facility (PAF) on the outskirts of Binghamton, New York. The fierce wind that swept the open lowlands around the city made us hunch our shoulders and raise our collars against its bitter chill. Back home in the Adirondacks the temperature was colder, but we shivered more here with few forests or mountains to break the force of the wind. Even so, the university campus upon which the PAF stood was bustling with students and faculty despite the weather.

We had come to the PAF to photograph the ancient, fluted point that Daniel Seib's team collected at Tupper Lake in 2007. It was housed in a building named, plainly enough, "Science 1," and when we arrived on campus the director, Laurie Miroff, welcomed us inside.

Laurie guided us through a maze of laboratories, classrooms, and corridors plastered with scientific research posters on our way to the basement. "It's kind of fitting," she quipped with a smile, "that our lab is located underground." Departments of geology or archaeology are often assigned to such subterranean settings on university campuses, but who needs windows when the objects of most interest to you lie belowground anyway?

A door opened and we entered a long room jammed with cabinets, stacks of cardboard storage boxes, and work tables well stocked with research equipment. There Laurie introduced us to Claire Horn and Brian Grills, two experts in the collection and care of items that are unearthed by archaeologists at proposed construction sites or other places where artifact-laden deposits are uncovered. A similar situation had led PAF investigators to survey the shores of Tupper Lake when the fallen water levels exposed them.

Claire, the lab director, showed us a cardboard box full of labeled pouches that resembled clear sandwich bags. "These are from Tupper Lake," she said, and placed the collection on a table beside us. Moments later, the main object of our visit was resting gently on the palm of my hand.

The ancient fluted point reflected the warm glow of a tabletop lamp from dozens of expertly crafted facets. It looked and felt as light as a leaf. The fresh-looking textures and edges made it difficult for me to believe that this was one of the oldest definitive signs of human presence in the Adirondacks. As I slowly turned it over and over in my hand, the combination of finely sculpted features, both wafer-thin and razor-sharp, was astonishing. How could a thing of such fierce but delicate beauty have survived a powerful thrust into a large animal?

It might not have survived undamaged, in fact, because a small section of the base of the point had somehow snapped off, although when and how the break occurred may forever remain a mystery. Perhaps it broke while penetrating the flesh and bone of a caribou. Perhaps it had been retrieved when the carcass was butchered, thenceforth to be repurposed as a knife, scraper, or even a souvenir from a memorable hunt.

While Kary and I took turns photographing the point, Claire searched among bags of pottery fragments and quartz flakes for other items of interest. "This one is nice," she said, holding up what at first seemed to be a pale nacho chip as large as her hand.

It was a triangular shard of quartzite, a kind of stone that is abundant in streambeds and glacial gravel in the Adirondack uplands and therefore would not require importation from the lowlands where chert was quarried. It looked relatively coarse-grained and therefore more difficult to work into fine features such as those found on most chert points, but the edges were surprisingly sharp.

We speculated about how the quartzite tool was used, maybe as a knife or a kind of hunting point that archaeologists call a "broadspear." Then a smaller, darker object emerged from the box, one that proved to be as remarkable as the fluted point we had come to see.

"I'm not sure what to call this," Claire said, and held up a chert point whose photograph I remembered seeing in Seib's report. "Let's check with Brian. He's the senior project director and should be able to identify it."

I soon realized that my intense focus on uncovering the earliest signs of human presence had caused me to overlook an exceptionally rare find of another sort. While not the oldest object in the collection, this ancient point was still remarkable in its own right. It was half the length of the fluted point, thickest at the center in cross-section, and shaped like a pyramid. Jagged features at the bottom, however, had perplexed me when I saw the photo in Seib's report, and I had not paid close attention to the technical description of it.

Brian emerged from his office holding a well-worn guide to projectile point identification. "Yes, I think that's it," he said in a hushed tone. "Kirk Corner-notched style, but with one of the notches broken off. It probably dates back to the Early Archaic period."

My heart skipped a beat. In researching this book I had learned that Paleoindian peoples left few traces of their presence in the Northeast, but signs of those who came soon after them are even more elusive. For years, the scarcity of Early Archaic sites led many archaeologists to believe that people had completely abandoned the Northeast between about 9,000 and 8,000 years ago.

If Brian's identification is correct, then this broken point makes a key contribution to our understanding of Adirondack history. It shows that people did indeed live in the uplands even during a little-known chapter in the saga of the region, a gap in the record not unlike the so-called Dark Ages of Europe. It also supports a growing consensus among experts that the shortage of published reports about this enigmatic time period does not mean that people didn't live in the Northeast then. Rather, it is mainly due to the limitations of archaeological work thus far and the low-impact lifestyles of those long-vanished inhabitants of the region.

"Nobody knows exactly why Early Archaic sites seem to be so rare," Brian explained. "Maybe it's because the populations were small, or maybe they didn't stay put long enough to leave many artifacts for us to find in any given place." It is also possible that the archaeological community has simply confused too many Early Archaic sites with earlier or later ones. That is especially likely where well-preserved layers and suitable materials such as charcoal, wood, or bone are not available for radiocarbon dating.

Above: A quartzite implement of uncertain age and function from the PAF collection that was found by Dan Seib's team at Tupper Lake.

Opposite: A probable Early Archaic projectile point that was found at Tupper Lake.

Kary and I thanked our hosts, bade them farewell, and returned to the car. A long drive home to the mountains awaited us, but we had much to contemplate along the way. What could we learn about life in the Adirondacks 8,000 years ago with so few artifacts to guide us? To answer that question, we must be prepared to savor the unknown along with fragments of insight that can be gleaned from discoveries elsewhere.

For a scientist such as myself it can be frustrating to try to reconstruct such deep history from incomplete data. It feels like trying to interpret a puzzle that has lost too many of its pieces to make a fully recognizable picture. Moreover, I'm not a formally trained archaeologist, so I'm not fully conversant with many of the peer-reviewed publications in that field. Instead, I must rely on interviews with experts from reputable institutions like the PAF, literature searches, and my background in reconstructing ancient environments to help fill gaps in the story.

On the other hand, when little hard evidence is available to refute or support an idea, even professional archaeologists can sometimes be tempted to indulge in reasonable speculation. Let's indulge in some of it ourselves now while keeping in mind that our guesswork is tentative. We may not get many firm answers, but we can have fun trying to do so with these three questions: Why are there so few Early Archaic sites? How did people procure food back then? And what technologies might people have used that are not well represented in the archaeological record?

Sediment cores from lakes of the uplands might help shed light on the first question. A core from Wolf Lake that was analyzed by Queen's University student Gladys Kong contained microscopic particles of forest fire charcoal, especially in the layers that were deposited between 13,000 and 8,000 years ago. Another student, Rebecca Hansford, examined pollen grains in the core and found that warm, dry climatic conditions also encouraged the growth of hardy pines between 11,000 and 8,000 years ago. Basal ages of cores from Adirondack peat bogs also show that the surface levels of many upland water bodies were lower than today during the Early Archaic period, and therein lies a clue.

What if early Adirondackers spent most of their time close to rivers and lakes where food and water were plentiful, as later peoples were often known to have done? As formerly warm, dry climates became wetter in more recent millennia, shoreline erosion and lake expansion could have destroyed or drowned many of the places they once occupied, making signs of their presence more difficult to find.

Basic aspects of the quest for food can also provide clues to the second question. It is unlikely that anything edible would have been overlooked in those days, so hunting of large animals need not always have played a dominant role in feeding a migratory band. Perhaps these Adirondackers focused instead on whatever could be readily snared, netted, hooked, or plucked. In that case, the scarcity of distinctive projectile points that identify Early Archaic sites might have less to do with small population sizes and more to do with lifestyles.

The implements of daily life used by those people could also make ancient sites more difficult to find if the most diagnostic tools were not made from durable stone. Properly crafted bone or fire-hardened wood can pierce hide and flesh as easily as sharpened chert. Might the scarcity of Early Archaic points reflect a greater reliance on more perishable materials that have since decayed away?

Fire-making was another aspect of daily life that can be difficult to reconstruct with certainty. Before the invention of matches and lighters, people had fewer options for starting a fire. They could strike a spark using a mineral such as pyrite, or they could use friction to generate heat directly from wood. Ancient pyrite strikers are rarely found in the Northeast, which suggests that friction was a more common method of fire-making in this heavily forested region.

There were two main options for making friction-fire. One of them involved a bow strung with a loose leather thong that was used to spin a vertical wooden spindle back and forth atop a small board. Intense friction where the whirling spindle met the board produced sawdust, blackened it, and finally yielded an ember that could be blown into a flame. Another method was to simply twirl a wooden spindle with one's bare hands. I never mastered it myself, but I have winced at the sight of the blisters one of my students developed before demonstrating the technique for me. We can only guess which methods were most commonly used long ago because wood, leather thongs, and even pyrite can decay or corrode until nothing of them is left for archaeologists to find.

A recent study by scientists at Simon Fraser University suggests that many early people might have been as intimidated by the rigors of fire-making as I was. Their comparisons of fire-making traditions among hunting-gathering communities worldwide showed that mastery of the craft was often restricted to just a few skilled individuals. Borrowing embers from neighbors and carrying them from place to place in bundles of smoldering tinder were also common practices because they were easier than producing fire from scratch. If early Adirondackers depended upon a small cadre of fire-making experts in their midst, then they would have left few fire kits behind to begin with.

We might also surmise that such specialization could have extended to stone tools, as well, particularly fluted points, which are exceptionally difficult to make. The beautifully flaked point from Tupper Lake is clearly the work of an expert, but reasons for the existence of the flutes remain unclear. While flutes were useful for attaching a point to a shaft and could perhaps make a point more resilient on impact, the ancient style was abandoned about 10,000 years ago.

Well, what if an elite caste of highly trained flint-knappers harnessed the natural conservatism of people whose survival often depended upon following well-established cultural traditions? We can then imagine such an expert proclaiming, "Hunting points only work properly if they are made in this demanding manner, so don't bother trying to make your own. You'll be better off relying on a master craftsman like me for your supply." Maybe a haunch of caribou seemed like a reasonable trade for a pouch full of specialist-endorsed points. But then why did people of the

Early Archaic period stop making fluted points altogether after keeping the tradition alive after so many millennia?

These are just a few of the questions we might never have definitive answers for. However, one such question is worth considering further here because it is relevant to the false claim of absence and is also an ongoing topic of inquiry among professional archaeologists. That question is not so much *whether* people lived in places like the Adirondack uplands, but *when*?

An ancient projectile point, the remains of a campfire, or small flakes of chert can be conclusive evidence of human presence in ages past. However, determining exactly when people stayed in any given place requires more information. Are there any tools or other objects that can clearly represent a particular time of year?

Spear and dart points are of little help in identifying seasonality because hunting in general could occur throughout the year. The same is true of the remains of campfires and cooking hearths.

Would fishhooks, net-sinker weights, or harpoons indicate warm-weather fishing? Not necessarily. Ice-fishing is a long-standing tradition in the North Country, and even nets can be deployed through holes in lake or river ice.

A dugout canoe or other type of watercraft would only be used in warmer months, so it could help narrow down the range of occupation seasons. As we will see in a later chapter, a few dugouts have been found in the Adirondacks, but they are relatively recent in age. Wood doesn't typically last long in the damp, fungus-rich Northeast, and boats that remain on lake bottoms for thousands of years can become hidden beyond reach by thick shrouds of mud.

Domesticated crops were not widespread until the last millennium or so, but even then their presence in an archaeological site might not necessarily indicate a warm season because the foods they yielded could be stored for later consumption. Likewise, wild berries or nuts could be dried, ground, or preserved for use throughout the year.

On the other hand, some foods can indeed help identify the season in which they were used. Distinctive mineralized deposits that formed on the bones of female passenger pigeons during egg-laying season could document pigeon hunting in spring. The remains of a small fawn would represent early summer, and a deer skull with large antlers attached would indicate autumn or early winter. But bones, like wood, are more likely to decay or be gnawed to pieces by rodents than to be preserved.

In short, we are again left with partially informed speculation to reconstruct the seasonal activities of early Adirondackers. Fortunately, we are not alone in that situation. Professional scholars also indulge in this kind of speculation, and we can use their work to help us imagine life in the uplands 8,000 years ago.

The scientists who have studied ancient hunting fences beneath Lake Huron have used their knowledge of the submerged terrain, local stone implements, and information from more recent hunting-gathering cultures elsewhere to infer a seasonal pattern of activity for early peoples there. For instance, the design of the fences offered a clue. They seemed to be oriented in such a way as to intercept caribou who were traveling northward, which would limit that form of hunting activity to spring when the herds' northward migration would have taken place.

In autumn, early Adirondackers probably ranged widely within their home territories to obtain hides and meat for the coming winter. During the colder months they might have sought more rugged upland terrain that provided shelter from harsh winds and subsisted there on hunting and ice-fishing as well as stored foods. Alternatively, they might have moved downhill to the Saint Lawrence estuary instead in order to ice-fish, hunt, and gather shellfish.

We can also infer reasons for people to congregate at certain times of year. Migratory fish such as alewives or shad might have been plentiful where Adirondack rivers met Lake Champlain, the Saint Lawrence River, or the Champlain Sea before them. Early peoples elsewhere in the Northeast exploited them, and the same species would probably have been available to early Adirondackers, as well. Spring in particular would have been a likely time for Adirondackers to gather at the mouths of certain rivers in the lowlands to intercept such fish.

The "come-together-for-fishing, disperse-later" hypothesis is supported by the presence of large numbers of stone items found in lowland sites such as the Ausable River farm, which might represent large gatherings of people. This suggests that people congregated in the lowlands for fishing or other activities, while less productive archaeological sites in the uplands might indicate smaller, more mobile groups during other seasons.

Then again, the scarcity of sites in the uplands could simply reflect the scarcity of research there. Springtime migration of white suckers and smelt in streams of the uplands could have led people to congregate there for fishing, as well. And as we shall see in the next chapter, several lakes in the uplands have yielded large numbers and varieties of stone tools that could reflect large gatherings there.

Despite all of these uncertainties, we can nonetheless be certain that people lived in the Adirondacks even during a period that many scholars once believed had left most of the Northeast uninhabited. Most details of those distant times will remain forever shrouded in mystery, but the evidence that we do have demonstrates the long-term presence of early Adirondackers and their enduring connections to the land.

We have already seen that rivers played a vital role in shaping and sustaining life on this upland landscape. But what about the still waters? Let's now examine some of the equally enduring connections between people and the lakes of the Adirondacks.

PEOPLE OF THE LAKES
(5,000 Years Ago)

Three thousand summers have passed since we last met the residents of the lakeside knoll, bringing us to another warm day 5,000 years before our own time. The knoll and lake look much the same, but the colors and textures of the mountain and the land around it have changed again. Enormous white pines with soft, bushy boughs still sigh along the crests of ridges, and spike-topped crowns of spruces and firs still darken the low areas. However, the forest mosaic is now more enriched with the lighter greens of leafy maples, birches, oaks, hickories, and beeches.

As before, we are not alone. Another young family, all clad in supple buckskin garments, has joined us on the knoll where we can watch them unnoticed. Mother and Daughter are singing as they gather sweet treats from blueberry bushes who carpet the ground beneath a grove of scrawny red pines. Father and Son sit on the breezy rim of the knoll and take turns flaking a projectile point of gray chert as their ancestors did on this very spot long ago. Their project, however, is more successful than the earlier one that ended with the snapping of a valuable piece of tool-stone. There is no troublesome fluted groove to contend with in this later style of dart point, and the boy has made good progress in sharpening the edges and rounding up the lashing notches near the base that his father has started for him. The antler tool clicks, small flakes of chert drop from his padded palm to the sandy ground, and Father's soft words of encouragement punctuate the melodies coming from the berry patch behind them.

Academic scholars in our own time will attempt to know people such as these by analyzing the stone implements and food remnants they lost or discarded. The styles of their projectile points will be given names like "Brewerton" and "Vergennes," and the cultures that produced them will be saddled with a term that can evoke images of primitiveness or backwardness. Members of the so-called "Archaic" cultures that coevolved with forest ecosystems of the Northeast between 9,000 and 3,000 years ago were not backward obsoletes as the label might seem to imply, but highly skilled, adaptable, and successful occupants of the land. Their lifestyles lasted several times longer than the Roman Empire and twice as long as the civilization of ancient Egypt, but they had far less impact on the land than the aforementioned cultures, a sign that they were more a part of their ecosystems than exploiters of them. As a result, they left relatively little physical evidence for modern archaeologists to interpret, but in comparison to the rare traces of their predecessors the physical legacies of these Adirondackers, especially those who lived during the last half of the Archaic period, are widespread in the uplands.

Let's take a closer look at some of those legacies now.

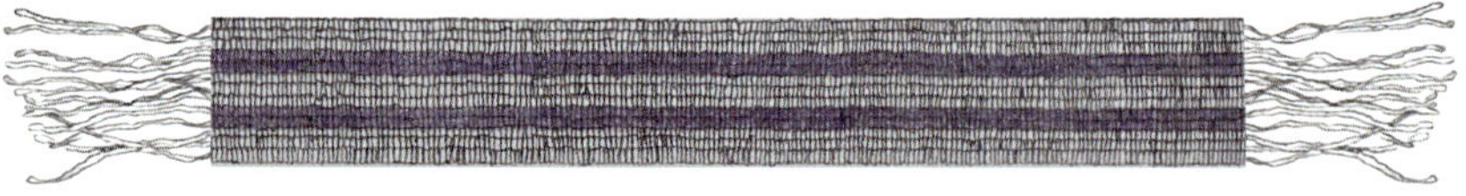

Several years ago, a young couple invited me to join them on an excursion to the southern uplands in order to introduce me to a family collection. We drove two hours on winding roads from Paul Smith's College through the heart of the Adirondacks to the Fulton Chain of numbered lakes near Old Forge in the Black River watershed. The long human presence in that sector of the uplands left behind many ancient items that have been gathered up by collectors such as my hosts' family during the last century or so, and in order to help prevent further unauthorized removal I will not specify exact locations here.

Susan Winchell-Sweeney, an archaeologist at the New York State Museum, would later explain to me that, if no digging is done and the items were already exposed in plain sight, the context of potentially datable layers is not necessarily much disturbed by surface-gathering on private property. Objects made of stone and ceramic can easily outlast us as they did their first owners, and if properly curated they can continue to inform and inspire generations to come. However, there is still a risk of permanent loss to posterity if the finds are not properly curated. People who gather ancient items often stash them away without descriptive labels in shoeboxes, drawers, or attics from which they are eventually dispersed or discarded when the finders die or move away.

The risk of losing valuable archaeological information can be lessened if accurate site descriptions, photos, or other records of the discoveries are shared with established keepers of history. "We try to reassure collectors," Winchell-Sweeney continued, "that we're not out to take anything away from them. All we ask is that they let us know when and where they find something, or at least that they label the items clearly for the sake of people who might inherit or study them later on."

After visiting the family's favorite collecting site at one of the lakes, my companions showed me some of the stone implements that had been discovered there over the years. They were beautiful to look at for their own sakes, but they also carried with them stories that I was just beginning to learn how to decipher. For example, the projectile points lacked flutes and basal grooves, which showed that they were less ancient than the Paleoindian point found at Tupper Lake. Instead, their bases were notched to help with lashing them to the tips of darts or spears, and most of their shapes reflected later cultural styles that were widespread in the Northeast between 6,000 and 4,000 years ago.

The materials from which the points were made were also informative. Most were flaked from imported chert, but one smooth, slender point was fashioned from brick-red slate, probably from the Champlain Valley or elsewhere in New England. These messengers from the past tell us that early Adirondackers did not live in total isolation but were connected to far-ranging networks of people and ecosystems. Then as now, the uplands were not backwaters but active cultural crossroads.

The Fulton Chain region has been worked over by collectors for quite some time, as is also evidenced by the large number of ancient implements archived at the Adirondack Experience museum in Blue Mountain Lake. They include various types of projectile point, slender drills made of flaked stone, scrapers, cutting tools, and a cylindrical bannerstone that was once attached to the shaft of an atlatl. All were crafted about 5,000 years ago.

The museum's collections also include younger stone implements and pottery fragments from the Fulton Chain ranging in age from 2,000 to

Above: A tiny flake of imported chert found at a state-operated campsite on Lower Saranac Lake. Like many others found throughout the Adirondacks, it attests to a widespread Indigenous presence in the uplands through the ages.

Opposite: Two intact projectile points, a broken point, and a teardrop-shaped scraper from the Fulton Chain.

400 years old. Similar items of various ages have also been documented at Algonquin Lake in the southeastern uplands, in the central Adirondacks at Rich Lake, and farther north at Tupper Lake and Lower Saint Regis Lake. Together, these collections represent a legacy of repeated connections to individual upland water bodies over thousands of years.

What was life like for Adirondackers 5,000 years ago? We can infer some things about those long-ago lifestyles from archaeological excavations, but another approach is simpler. Here is what a student and I learned during an informal historical experiment on a lovely lake in the northern uplands.

It was a brisk but sunny morning in April when Ken and I launched my carbon fiber canoe on a marshy stream that would lead us to Middle Saranac Lake. Our mission was twofold: to enjoy the first outing of spring and to experience this familiar place with fresh eyes. Ken was a student who had become interested in local environmental history through my classes at Paul Smith's College and was keen to explore the lake and its shoreline with people of the past in mind.

Our approach was to try to envision living here 5,000 years ago. We knew that our imaginations had been shaped by novels and films and filtered through a macho mythos of braving the primal wilderness alone with minimal equipment. However, we hoped to bypass such stereotypes with the aid of a real-world setting in order to gain a better understanding of what it meant to be an Adirondacker in ages past.

From the scientific literature we knew that the climate, forests, and wildlife of the region today are roughly similar to those of 5,000 years ago. Although the arrival of Euro-Americans devastated many species during the last few centuries, beavers have made a comeback and moose are recovering. The once-abundant passenger pigeons are gone forever, but turkeys, grouse, ducks, and geese are still here, and deer are relatively abundant. Artificially introduced bass and pike have replaced most of the trout in the lakes, but the overall biomass of fish seems to be about the same. If we could ignore the presence of carbon fiber canoes and the drone of distant cars in the background, then Middle Saranac Lake would be a good location for our project.

When we first considered the kinds of food that were theoretically available to people of the uplands 5,000 years ago, it seemed like a rich cornucopia. Excavations of Late Archaic refuse piles elsewhere in the Northeast have yielded signs of meals based on deer, bears, hares, geese, bullheads, turkeys, grouse, and turtles as well as acorns, cattail tubers, goosefoot seeds, assorted greens, berries, and other plant foods. Such a list may evoke images of a generously stocked buffet, but we soon realized that it said nothing about the availability of those foods in practical terms. Here on the lake we were already growing hungry with no meal in sight. Where was breakfast?

Our first stop was a tiny pine-clad island near the center of the lake. After pulling the canoe ashore, we clambered up the steep slope to the top. We found no obviously edible plants there other than the trees themselves. Early residents of the Northeast sometimes sliced, roasted, and processed the inner bark of conifers into thin strips or fine powder to yield a pungent famine food. The high-fiber treat can be tasty when properly seasoned, but it is better at filling the belly than sustaining life. Ken and I were not that desperate yet, but we quickly noticed that we had arrived too early in the season for berries or greens. It was clear that we needed to find some animal-based food somewhere, somehow.

While circling the rocky shoreline, we suddenly stumbled upon a potential meal. A pair of Canada geese dashed noisily into the lake ahead of us, leaving behind a cache of cream-colored eggs in a plush nest lined with fluffy down.

"I told you so," Ken said. I had been skeptical that we'd have an easy time, and this lucky find seemed to support Ken's more optimistic outlook.

"OK," I replied. "Let's imagine that we cook and eat these eggs. That would finish off the nest, and we could raid every other nest on the lake within a day or two. Then what?"

Could we hunt the geese? Sure, in theory. But it's not necessarily easy to get close enough to wary waterfowl to grab or hit them with something. The distressed parents who honked at us from offshore would not let us do so now, and there were no other waterbirds in sight. Even if we did manage to catch this pair, there would be no more nesting geese here for us later on.

Now another problem came to mind. We would probably not be the only people living here. There could also be grandparents, parents, siblings, spouses, and kids to consider, all of whom would need to eat, as well. Who would really have had those few goose eggs for breakfast today?

"It looks like we'll need to hunt up some deer instead," Ken suggested.

"All right," I replied. "Where?"

Foraging in the woods is not like entering a supermarket, finding the meat section, and ordering a preferred cut whenever hunger demands it. Neither of us had seen many deer in the forest during our hikes in the past, mainly encountering them along roadsides while driving at night. None were visible now, and if we were to wander through the surrounding forest in search of them, they would likely detect us first and flee. Hunters today don't normally sneak up on deer like they often do in the movies but are more likely to sit in wait for them where the animals are already known to travel or browse. Even then, it's not uncommon to come home empty-handed. And this is with modern rifles or high-tech hunting bows for armaments, not atlatls, and with plenty of grocery store food waiting at home as backup.

Ken was young, single, and ready to handle hunger and uncertainty as an individual, but as an elder I was more concerned about potential family responsibilities. We could wander for days in pursuit of a deer while our loved ones slowly starved. As we were learning firsthand, hunting was not a mere pleasure sport in those days but a matter of survival that was supplemented by more generalized foraging. Most of the foods that early Adirondackers could have relied upon were not always available at all locations or on demand, so finding them required a strong working knowledge of the landscape and the plants and animals who inhabited it.

Where sugar maples grew in quantity, the waning days of winter could send sap flowing into bark containers where it would freeze overnight, leaving sweet liquid beneath the ice. Spring once saw passenger pigeons congregate in huge numbers for the nesting season. Turtles crawled ashore to lay their eggs then, and white suckers and smelt swarmed up certain streams to spawn then, too. Half a dozen kinds of berry ripened sequentially throughout the summer. Eels migrated downstream toward their mid-ocean breeding grounds in autumn. Cattail tubers and the rootstocks of water lilies who grew in the muddy shallows of lakes and wetlands were most nutritious in autumn after the growing season had filled them with starch. Acorns, hickory nuts, and beechnuts could only be harvested in certain sectors of forest at certain times of year, and bears, raccoons, and turtles could only be taken when they were not sleeping out of sight in winter.

As Ken and I absorbed all of this, another thing became increasingly clear. We might not really be able to support a family all by ourselves for very long, especially without the fancy sporting gear that is available to us today. Early Adirondackers had to work together, share what they were able to obtain, and follow the seasonal rhythms of the far-flung ecosystems they lived among. Rather than stay put and expect the world to sustain them as most of us do nowadays, they lived in portable or easily constructed shelters and learned how to find and gather food when and where it became available.

In early Adirondack communities, waiting for a lone hunter to come home with a buck on his shoulders would probably not have kept everyone fed consistently. People gathered what they could, when they could, rather than go to bed hungry. Almost anyone in the community could set snares and deadfall traps, pluck freshwater mussels or net minnows in the shallows, catch frogs and turtles, knock birds or porcupines down with stones or sticks, or gather herbs, seeds, and fruits where possible. Sharing food was crucial and being picky about one's diet was a rare luxury. Everything edible was likely to be eaten, be it a prime slice of tenderloin or squishy entrails.

In addition, hunting in groups was often a more effective way to capture large animals than solo stalking. Helpers could drive a moose or deer into a lake, river, or deep snow where the animal slowed down and became an easier target. In some cases the woods might be set afire or a ravine might be used to funnel prey toward waiting hunters. Hunters might also turn their dogs loose to locate, chase, and tree a bear or to help drive their quarry into the water for easier killing.

For early hunter-gatherers to live sustainably in the Adirondacks required a strong sense of community supported by family and culture. It also required a practical grasp of ecology, a field of study that western science has only recently rediscovered. Overexploiting any one place or wasting food could mean starvation, and harsh lessons that were learned long ago by trial and error were woven into stories, social norms, and the collective memory banks of traditional ecological knowledge.

Some academics who study this kind of lifestyle have called it "restricted wandering." It involved using an entire region as a home, with relatively small groups moving around in their own favored sections of watersheds with the familiarity we might have for our hometown. There were limits to how many persons the land could sustain, so protocols of respect for a group's territory must have existed, as well, though they might not always have been followed if conflict was in the cards. But for a band of early Adirondackers, saying "this place is our home" did not proclaim property ownership in the way it might today. The land was not a commodity that belonged to human beings any more than it belonged to the deer or geese. More often, the relationship worked the other way around.

The waters of the lakes and streams, the minerals of the rocks and soil, and the air that rustled the leaves of the forests produced and sustained the web of life within which the human residents were equally enmeshed. These Adirondackers were elementally bound to the world in which they lived, and they knew it. They were truly people of the lakes.

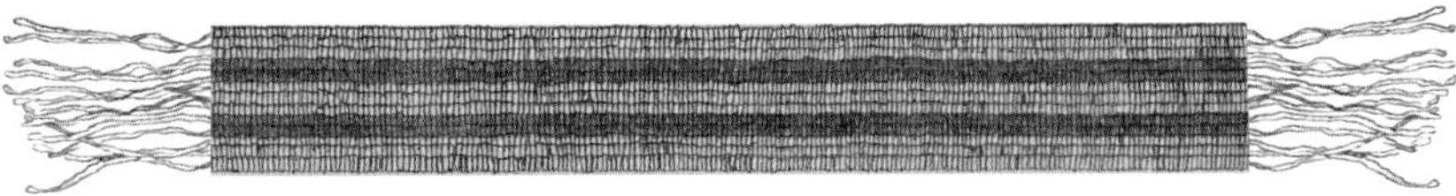

Let us return to the knoll now. Summer and fall have passed since we encountered the family of uplanders. It is midwinter, and the lake has become a solid plain of ice.

Smoke rises from the tops of several conical, bark-covered lodges that nestle on the lee side of the knoll where they are sheltered from winds that sweep across the frozen lake. These people well know that ridges and hills can offer protection from bone-chilling blasts that more freely torment the flatter, open landscapes of the surrounding lowlands. The uneven upland terrain also produces a range of snow depths that can serve various purposes, from easier travel on ice-paved waterways to thick drifts that can bog a moose down and make for easier capture.

Winter is the traditional time for storytelling, and this morning Grandmother is sharing her memories and life lessons with the children. They have gathered around her beside the warmly glowing hearth at the center of the lodge. We can only guess at her words, but it is likely that she will tell the strange story of the hemlocks at some point.

When Grandmother was a little girl, she used to play with the tiny cones that littered the floor of a hemlock grove on a small glacial hillock nearby. But by the time she became a young mother the dark green needles in the canopy had turned brown and fallen, leaving barren branches behind. Gaps opened where majestic trunks once stood, and dense thickets of maple and beech saplings crowded in beneath them. The loss of her childhood playground made her sad, but when a visitor reported that the same thing was happening to hemlocks all over the region, she and her family began to worry. Was it some sort of curse?

The children are speaking up now, and they seem to have something to say about that. They know the hill Grandmother is speaking of, and it is now their favorite place to gather raspberries. "Maybe it wasn't a curse after all," one little girl perhaps is saying. "I like it the way it is now!"

The sudden loss of the hemlocks was frightening at first, but when some forests across the Northeast became richer in nuts, berries, and browse as a result of it the deer, bears, wild turkeys, and other animals could presumably have become more abundant, too. So too, perhaps, might have the people. Archaeologists have noted that signs of human presence in the Northeast became more numerous about 5,000 years ago, and some suspect that a population boom in edible wildlife due to the mysterious hemlock collapse might have been a contributing factor. However, without knowing how many people really lived here at the time,

we may never know what, if any, effect the decline had on the people of the Adirondacks.

But it's time now to find something for dinner. The bears and raccoons are asleep in their dens at this time of year, and the turtles and frogs lie dormant in the muddy floors of the lakes and wetlands. The passenger pigeons and waterfowl have migrated south, the vegetable department will remain closed until spring, and the deer and moose have learned to avoid this inhabited knoll. But no worries. Fresh food awaits underfoot for those who know how to find it.

The wind has subsided, and the bright sun and clear blue sky invite people to enjoy the fresh air outside their dark, smoke-filled lodges. The natural cycles of the earth's spin and tilt have made climates somewhat milder in comparison to those of previous centuries, but one still needs to dress for cold at this time of year. The women have worked hard to provide everyone with buckskin garments trimmed with the soft furs of wolves, foxes, and fishers, and their moccasins of thick moose hide are insulated with dried moss and waterproofed with animal fat.

Everyone but the elders, who remain behind to tend the hearths, has now ventured out onto the frozen lake, which the wind swept clean of snow during the night. The youngest kids sprawl flat on their bellies, counting frosty white disks of methane that bubbled up from the bottom of the lake and became trapped within the glassy clear ice. Others run and slide as far as they can, giggling and shouting for their mothers to watch.

Half a dozen men and older boys stand a few paces apart and chip away at the ice with bone-tipped poles, gradually opening holes wide enough to pull fish through. They hunch over the holes, clutching lines of twisted dogbane fiber equipped with hooking devices that are embedded in bits of fat or entrails. The devices vary from curved hooks to needle-shaped gorges of bone, copper, or wood that can be yanked crosswise in a fish's gullet. Each one is lowered with the aid of a thumb-sized stone sinker with a small groove at the top for lashing to the main line. Soon the ice around the holes is littered with trout, and a pair of bold ravens has landed on the ice nearby in hopes of snatching scraps.

The sun is beginning to slide behind the mountain now, and it is time to return to the lodges. Any fish who are not roasted for dinner or fed to the dogs tonight can be hung from the lodge poles for drying and smoking. An uncle who has just arrived from an adjacent section of the watershed has also brought with him a special treat, a plump beaver whom he speared through a hole in the ice as the animal swam out of their lodge. The meat will be delicious, the sharp front teeth will become woodworking tools, and the fur will help make a garment soft and warm. Sticky, odiferous castoreum from the beaver's scent glands will also help bind fletching to atlatl darts.

Now more stories begin as countless stars sparkle above the mountain and orange firelight sets the open smoke holes of the lodges aglow. We can imagine that Grandfather is reciting that same old tale about a deer whom he chased onto a frozen lake as a youth, wearing clawlike bone cleats that helped him keep his footing while the deer sprawled on the slick ice, helpless to resist his spear thrust. He gets as far as "You'd never guess it, but I learned that trick by watching . . ." when another voice interrupts him.

"The wolves!" Grandmother has heard this tale often enough to know the ending by heart. The whole family erupts with laughter.

Now Uncle speaks up to tell stories of Glooskap, the magical hero of Wabanaki traditions. In one account, Glooskap summons a gigantic primordial beaver, strokes the monster gently, and shrinks him down to his present size. Five millennia in the future, archaeologists will hear the same account and wonder if it echoes folk memories of the deglacial days when giant Castoroides roamed the lake country.

As these Adirondackers snuggle together beneath cozy fur robes for the night, their bellies are not rumbling with hunger. Nonetheless, the monotony of the winter diet is beginning to wear on them. Tonight, the elders may dream of fresh greens from summers past while the youngsters dream of the sweet strawberries they will gather when spring returns.

In a few more months it will once again be easy to dip-net suckers who swim up a nearby stream to spawn, to spread trotlines for bullheads at night, and to hang weighted gill nets from floating chunks of driftwood to ensnare lake trout. Beavers and muskrats will be easier to surprise again without the protective cover of ice. Waterfowl will return, and frogs and turtles will haunt the shallows once more. But regardless of the season or the foods that sustain them at any given moment, these Adirondackers are people of the lakes, and the physical traces of their presence here will whisper softly to us if we listen carefully from our own position farther along on the deep waters of time.

CROPS AND CLAY
(600 Years Ago)

The Adirondacks are ablaze with the vibrant colors of autumn on this crisp October day six hundred years before the twenty-first century. The blueberry bushes beneath the pines atop the knoll are blushing scarlet, and the forested slopes of the mountain are daubed with bright splashes of birch yellow, maple red, and beech bronze. A chilly breeze pushes ragged waves against the rocks at the foot of the knoll where a family of four have gathered, all clad in smoke-scented buckskin and facing into the wind.

Mother sits on a cattail mat beside ten-year-old Daughter, who is struggling to shape a small clay pot properly. Father and ten-year-old Son are finishing up a pair of stone axes that they plan to use for making a dugout canoe from the trunk of a cedar tree whom the wind and waves have toppled on the water's edge. Now Mother rises to her feet and points toward the opposite shore where the outlet river exits the lake. Visitors are coming.

As the wind pushes the guests nearer, their birchbark canoe comes into clearer focus. Its seams, gunwales, and high pointed bow wear decorative trims of blackened conifer pitch sealant and spruce root lashings. A loon calls from across the lake and is immediately answered by an equally haunting call from the canoe. It is an improvised song coming from the man who is steering from the stern. The melody is clear, but distance makes most of the lyrics difficult to hear from shore, all except for one musically extended word.

"Mdawi-i-i-i-l HAH!"

The man repeats it again and points to the bird, who dives beneath the surface as if surprised to hear her name sung aloud. The language of the song identifies the singer as a member of a diverse group of Algonquian nations who have long occupied much of the northeastern sector of the continent. These visitors belong to the Abenaki nation whose territory stretches eastward across Lake Champlain and the Green Mountains, beyond which related peoples occupy much of what will later be called Maine and the Canadian Maritimes.

The design of the boat also helps identify the visitors because the bark canoes used by local Iroquoian peoples such as the Haudenosaunee, whose ancestors arrived in the Adirondack region just a few centuries prior, are normally made of elm, not birch. However, the Iroquoian family members on the knoll already know who these travelers are. The same group has come here every autumn for the last several years and have become familiar acquaintances. This Abenaki family is coming to pay their respects while their companions erect conical, bark-covered lodges on the upwind shore.

Father steps forward to help ease the delicate craft against the rocks, extending a welcoming hand to a boy and girl who are seated in the center. The children clamber ashore, greet Daughter and Son, and scurry off with them to a cluster of domed bark homes on the lee side of the knoll. A woman in the bow steps ashore next and embraces Mother while both men haul the canoe beyond the reach of the waves. Everyone is smiling, but there is little conversation going on; they are using hand signs and simplified vocabulary from two distinct languages.

The visitors speak in a manner that could still be understood among Wabanaki peoples of the Northeast in our own times, while the occupants of the knoll speak a version of Haudenosaunee language. There are other cultural differences, too. Men generally play more central roles in patrilineal Wabanaki society, while women have more significant governing influence in matrilineal Haudenosaunee society. At the time of this visit both cultural groups have considered the uplands to be part of their territory for generations. Their encounters with one another have produced variable degrees of friction in the North Country, sometimes even violence, but on this particular lake the interactions have thus far been respectful and pleasant.

As the two couples approach the hosts' lodge, other residents of the knoll smile or nod in greeting. A huge, round-bottomed clay pot of fish stew awaits them inside, suspended by a heavy cord over the central hearth. The bubbling stew has been thickened with maize meal and dotted with fresh cranberries from a nearby wetland. The visitors and their hosts sit beside the hearth on cattail mats, smoke fragrant tobacco in elbow-shaped pipes of baked clay, and savor the hot meal that Grandmother scoops into wooden bowls for them.

The Abenaki family have brought gifts as tokens of appreciation for permission to share the watershed and for the meal. The woman unwraps a lump of moist marine clay from the Saint Lawrence Valley that Mother will appreciate when, inevitably, someone breaks the current cooking pot. The man opens a leather pouch and shakes out a dozen sharp flakes of chert from the Champlain Valley that Father can shape into cutting tools and arrowheads. Arrows may strike with less force than heavier atlatl-driven darts, but in recent centuries their faster, straighter flights have made them the preferred weapon for killing prey or, on occasion, unfriendly human beings. Not to be outdone, Father gives the man a plush, chocolate-colored mink pelt while Mother places a handful of rolled copper beads from the upper Great Lakes in her friend's lap.

Soon the visitors will rejoin their companions on the opposite shore of the lake. They have spent the summer in the northern uplands and will soon be on their way south. Tomorrow they will paddle to the farthest lake on this chain, then make overland carries to other watersheds before joining relatives who traditionally overwinter in the upper Hudson headwaters.

Meanwhile, their hosts on the knoll will remain on this lake to hunt, trap, and ice-fish through the winter. When spring returns, they will follow a well-worn footpath beside the outlet stream to the lowlands along the Saint Lawrence River. There they will join family and friends in a large, palisaded farming village and grow maize and other crops.

These are generally good times for Adirondackers, with new ways of being on the land that have been developed with input from multiple Indigenous cultures and that help make life less precarious. In our own time centuries later, Haudenosaunee descendants will still refer to these uplands as the "dish with one spoon," meaning that many peoples have managed to share the region with a minimum of intercultural strife.

Sadly, things will not remain that way for much longer, as we shall see in coming chapters. But for now the brighter sides of human nature are ascendant on this upland lake, and all seems right with the world.

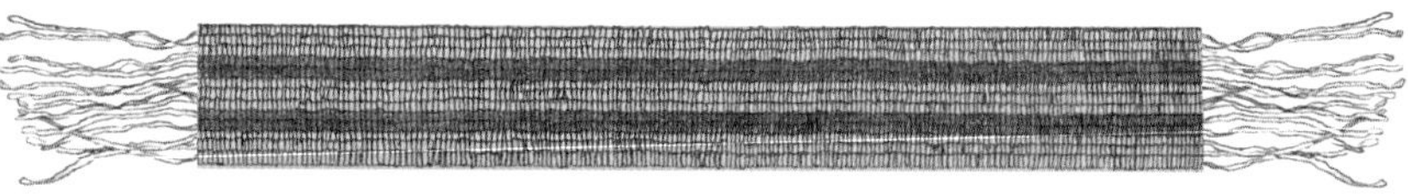

About a decade ago, I learned that SUNY Potsdam archaeologist Tim Messner was supervising an excavation at Long Lake in the central uplands. He was relatively new to the university but had already gained local fame for, among other things, constructing a dugout canoe with his students, using only fire and traditional stone tools to hollow out a huge white pine log, and then launching it successfully on the Raquette River. His manner of engaging students, peers, and the public alike was exemplary, and his interest in exploring the deep human history of the region was welcome, too. Few academic scholars had thus far devoted serious attention to documenting and celebrating the long Indigenous presence in the Adirondacks, and Tim was helping to change that.

Kary and I met Tim at a convenience store near the lake and followed him to a cluster of unoccupied cabins on the western shore. A local landowner had invited him to investigate the property, and the prospects for exciting discoveries there seemed excellent. Other sites in the area were already known to yield projectile points and bits of pottery, but we would see none of that here. Instead, Tim led us to the edge of a square pit that his students had sunk into the lawn and pointed down to a layer of rusty brown lumps on the bottom.

"That's what we call FCR," he said triumphantly.

After explaining that "FCR" translates to "fire-cracked rock," Tim described how those plain-looking cobbles revealed an ancient human presence. To our untrained eyes the smooth, fist-sized pieces of quartzite resembled many others that abound in streambeds and on wave-washed beaches throughout the uplands. The trick to recognizing their significance was to notice their odd rusty colors, crumbly textures, and numerous cracks.

If you've spent much time in the woods you've probably seen what fire can do to cobbles like these where people have surrounded their campfires with stones. The heat vaporizes moisture in the fabric of the rock and oxidizes its iron-rich components to rust while also loosening the mineral grains and often breaking the cobbles where expanding steam wedges them open.

"I think it's a hearth," Tim said.

This was not a hearth in the sense of a big stone fireplace in a cabin, but a former firepit in which cobbles had been heated, perhaps as a way to warm the interior of a shelter without smoking it up, but more likely for use in cooking. Tim explained the process to us.

"Before pottery vessels were available, people used containers made of bark or hide. The cooking was done by heating small rocks and dropping them into the food." When I later tried the method at home, it took only three glowing cobbles to bring a large pot of water to a boil.

How old might the Long Lake hearth be? Tim said he couldn't be sure from the FCR itself, but small pieces of charcoal that lay scattered among them could be radiocarbon dated. All that was needed was a few hundred dollars to pay for the analysis. My own research on Adirondack lake history included funding from the National Science Foundation for this sort of analysis, so a pinch of charcoal was soon on its way to a dating lab. A month later we had our results. The hearth was roughly 1,000 years old.

In contrast to claims that the Adirondacks were uninhabited until recently, Messner's excavations demonstrated that people were cooking meals on the shore of Long Lake around the time when Norse seafolk were colonizing Iceland. Perhaps this particular hearth was left by a band of Algonquian ancestors similar to those whom we have imagined encountering on Lower Saint Regis Lake. There were, after all, enough stones in the hearth to cook a substantial amount of food.

And what might have been on the menu on that distant day beside the lake? Many of the peoples who lived in and around the Adirondacks a thousand years ago had recently begun to incorporate farming into their lifestyles. Therefore, dried maize might have been part of the cuisine at Long Lake. Although horticulture represents only a recent, relatively brief chapter in the region's history, its importance to the development of local lifestyles then cannot be denied. Let's now examine the stories of some of those crops upon which so many people depended in the past and still do today. Where did they come from, and when?

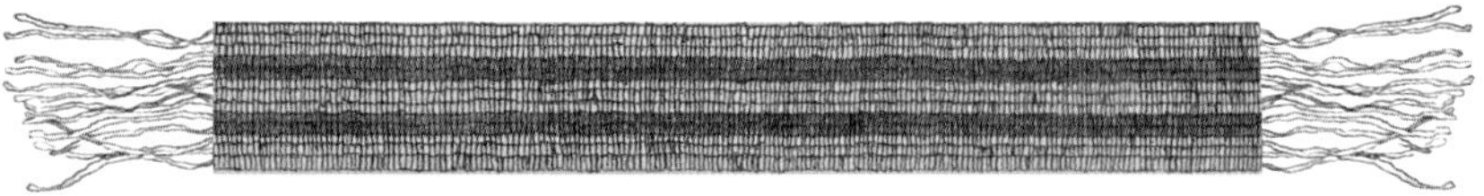

Many of us were taught in grade school that Indigenous people helped European settlers survive in the so-called New World by sharing the bounty of their crops, specifically maize, beans, and squash. Known as the "Three Sisters," those crops are often associated in the public's imagination with Indigenous cultures in general. However, the idea that the Three Sisters were always the basis of traditional lifestyles on this continent, meaning that regions unsuitable for intensive horticulture were therefore uninhabitable, is false. The real story is more nuanced and interesting.

John Hart, an archaeologist with the New York State Museum, studies the history of crop plants in the Northeast. During a recent visit to his office in Albany, I asked him about the legacy of the Three Sisters in the North Country.

According to Hart, standard versions of Indigenous history in New York State typically emphasize large, fortified farming villages of the kinds that were documented by early European visitors. The French explorer Jacques Cartier described extensive maize fields along the Saint Lawrence River in 1535, and Henry Hudson reported seeing similar fields beside his namesake river in 1609. But such portrayals don't fully reflect how people lived in the North Country through the ages. In fact, stockaded longhouse settlements and extensive cultivation only became common a few centuries before the European invasion.

"Even the most dedicated horticulturalists continued to hunt and fish and gather edible plants," Hart explained. "Many Indigenous farmers were seasonally mobile and used large family territories in the uplands too." In that sense, the lifeways of old had not really ended when intensive crop production began to support large Haudenosaunee settlements in the region. Horticulture was less central to much of Wabanaki culture, and it was common for Abenaki families to leave their crops untended while fishing, hunting, or foraging elsewhere.

As for the crops themselves, how old were the Three Sisters?

"The classic trio as a unit is relatively young," Hart explained, "and the crops weren't developed all at once. Squash was farmed here first, then maize. Beans were the last to arrive, probably less than a thousand years ago." A millennium may seem like a long time, but in the larger context of human history in the Adirondacks, cultivation of the Three Sisters represents less than one-tenth of the story.

Archaeologists have traditionally depended upon finding readily visible plant remains in their study sites to reconstruct the history of crop use, but a newer technique involves analyzing charred food residues on the interiors of cooking vessels. Even broken pieces of pottery can be scraped for examination under a microscope to reveal tiny lumps of glassy silica called phytoliths ("plant-stones"), which are naturally embedded in plant tissues as a defense against grazing animals. Experts can identify the remains of many plant species by their phytoliths and can also determine their ages through radiocarbon dating of the charred residues surrounding them.

Phytolith analyses demonstrate that, about 3,800 years ago, people in the Ohio Valley and Finger Lakes region were preparing meals from a wild native grass now known as "little barley." Other plant species also bore telltale signs of early domestication by then, such as larger and increasingly numerous seeds or widening geographic ranges. Among them were goosefoot, erect knotweed, and sumpweed. Although nobody was clearing and tending very large fields back then, the coevolution of plants and people was well under way in the forests and floodplains of the Northeast. Some or all of these wild or half-wild plants could have been helping to sustain people in the Adirondack region long before the Three Sisters arrived.

Squash was probably first domesticated in Central America, and people were growing the plant in Mexico 10,000 years ago. To better understand how squash fits into the story in the Northeast, however, we must first distinguish between squashes and gourds. Both belong to the same genus, *Cucurbita*, but the ones we call gourds are hard-walled, often round, and usually too bitter for eating. Squashes are softer, tastier, and more irregular in shape; think summer squash, fleshy pumpkins, and the like. One might expect to find delicious squashes in use before gourds, but in some cases the actual sequence was the other way around.

Squash was being cultivated roughly 3,000 years ago in what is now known as Central New York, but gourds were present in the region long before that. Their seeds have been found among signs of human presence dating back 6,000 years in the Northeast. The seeds can be made palatable by leaching them in water and wood ash, a technique that Indigenous peoples similarly employed to make astringent acorns taste better. But gourd cultivation might not have been all about food at first.

During my visit, Hart explained that gourds can be used as tools, too. "If you dry them out they make good rattles and containers, and they could also have been used as floats for fishing nets." It seems that the oldest of the Three Sisters began her relationship with humans by serving as a multiuse resource.

Maize, the next Sister, was also first domesticated thousands of years ago in Central America, where she was known through the ages by various names including "ixiim" (Maya) and "cintli" (Aztec). When the crop eventually spread through the Caribbean Islands, the local Taíno people called her "mahis," from which we derive the name "maize." Europeans arriving on the continent added the name "corn," which is a European term for grains in general.

Phytoliths found on pottery fragments show that maize had reached Central New York by about 2,300 years ago. Initially, residents of the region might have considered corn to be "just another starchy grain" among the others, as Hart put it, and it was only during the last millennium or so that maize became a major crop in the North Country.

The human-driven evolution of maize was most likely the work of women, who were typically the primary caretakers of crops. One of the early varieties they developed was "Northern Flint," with slender cobs and variable kernels that could be white, yellow, speckled, or blue. The hulls of the kernels were harder, or "flintier," than those in sweet maize, which made for easier storage in dried form.

The varieties of maize that were grown in the Northeast were well suited to short growing seasons. Samuel de Champlain reported that Indigenous farmers along the Saint Lawrence River planted maize in May and harvested in September. Versions of what might be called "fast maize" were also well known. John Winthrop, a resident of Maine, reported in 1670 that "Indian Corne of a special kind, (is) said to ripen a month at least before other kinds." An account written by a Penobscot author in 1890 noted that local varieties of maize could mature within three months, and one especially cold-tolerant type of heirloom Abenaki flint corn from Vermont is said to mature within eighty days.

Such hardy local forms would have been suitable for use in the Adirondacks long ago, too. Despite the higher elevations, average temperatures in most of the uplands are only a few degrees cooler than in the adjacent lowlands, pockets of warmer microclimates are common, and fertile soils occur on floodplains and the remnants of former deglacial lakes.

As for the Northeast's late-arriving third Sister, Eurasia has long known its own kinds of domesticated bean, but the ones who originated in Central and South America belong to the genus *Phaseolus*. Since then,

American beans have diversified under cultivation into pinto, kidney, lima, black, and many other forms. They reached the Northeast sometime during the last millennium or so, but they did not become common here until about 700 years ago.

Some experts have suggested that beans displaced a previous crop sibling who had once represented a more important component of the maize diet. Wild rice, like beans, contains nutritional elements that maize lacks, and the remains of rice grains sometimes mingled with residues of maize in clay cooking vessels before beans arrived. However, beans lose less of their nutritional value when cooked, and their vines thrive among cornstalks that serve as natural beanpoles. Those traits might have contributed to the rapid spread of bean cultivation across the Northeast in recent centuries.

With little archaeological work being done thus far on crop history in the Adirondacks, we can't be sure how long or how often the Three Sisters were grown in the uplands long ago. However, we do have written accounts from the 1800s of Indigenous women growing maize at Indian Carry on Upper Saranac Lake and near Piseco Lake, so crops were clearly grown here to some extent as supplements to traditional hunting, fishing, and gathering in recent centuries, if not longer.

Ceramics themselves might contain clues in that regard, too. The presence of heavy, fragile pottery throughout the Adirondacks challenges claims that the high country was avoided by all but a few transients who crossed it quickly, as we shall soon see.

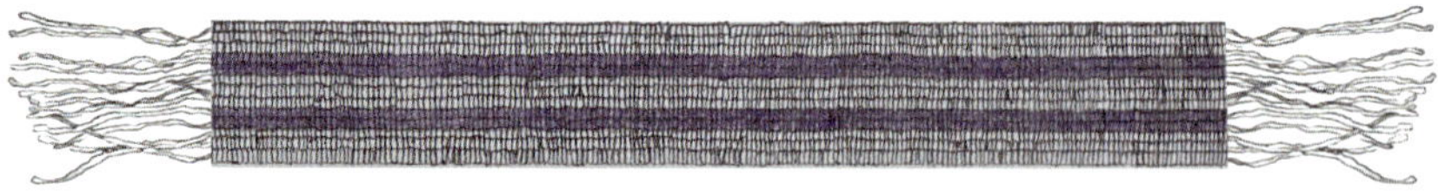

On a memorable autumn day during the 1970s, Saranac Lake resident Jim Bickford was savoring a quiet moment in the woods near his remote deer-hunting stand. Jim had discovered the spot after he and his companions followed a winding esker ridge deep into the forest near the base of

Saint Regis Mountain. There he found a beautiful grove of beeches and maples where a large rock outcrop made a perfect lookout platform.

With no deer in sight, Jim's mind had begun to wander.

"I saw a big crevice in the rock beside me," he recalled decades later. When he reached into it he was thinking that he was the first person ever to touch that spot.

He was mistaken.

"Then I felt the pieces under my hand." He pulled several of them out from among the dead leaves and other forest debris for a closer look. They were fragments of ancient pottery.

Jim Bickford isn't the only hunter to have found Indigenous ceramics in the Adirondacks. In fact, amateurs who roam the backcountry have found more such evidence of long human presence than professional archaeologists. Jim's fragments are now archived at the Six Nations Iroquois Cultural Center, but despite their historical significance they receive less attention from visitors than a more intact find that holds a place of honor in a glass case of its own.

During the 1940s, another hunter from Saranac Lake named Frank Woodruff noticed an overhanging ledge while exploring the base of a cliff near Silver Lake in the northeastern Adirondacks. The protective overhang provided ample room for sleeping spaces beneath it with a clear view of a wildlife-friendly marsh down below. But the greatest surprise lay in the dark interior of the shelter where Woodruff found a large clay pot lying on the rocky floor.

Fast-forward to 2019, when John Kahionhes Fadden showed me photos of his father, Ray, at the shelter with Woodruff. Woodruff had kept the pot in his attic but was preparing to retire in Florida and agreed to sell it to Ray, who installed it in the museum in Onchiota. The photos would later help me and my wife, Kary, locate the shelter and report its exact location to John for future reference.

"An expert who examined the pot for us said that it's about 300 to 500 years old," John explained, "and he was pretty sure it was made by Haudenosaunee people."

Above left: The rock shelter near Saint Regis Mountain where Jim Bickford found broken pieces of pottery. *Above right:* The rock shelter near Silver Lake where the large pot was discovered.

Opposite: Ray Tehanetorens Fadden holds the pot that was discovered in the Silver Lake rock shelter. PHOTOGRAPH BY JOHN KAHIONHES FADDEN

Below: Undeniable evidence that crops can be successfully grown in the Adirondack uplands. Saint Regis Mountain is in the background and the Tucker family's "Great Adirondack Corn Maze" field is in the foreground at 1,700 feet elevation. PHOTOGRAPH BY RICHARD TUCKER

Among the telltale features was a wide collar above the smooth round belly of the pot, which was decorated with raised points along the rim. The curvature of the pot's bottom suggested that it might have been nestled into the hot ash and embers of a firepit, stabilized with small rocks around the base, or suspended over flames by a sturdy cord around the collar.

The Silver Lake pot is a fascinating mix of information and mystery. Its fragility and unwieldy size suggest that it was used by someone who spent lots of time at the shelter rather than being hauled over the mountains while "only passing through." Traditional gender roles associated with Indigenous ceramics suggest that at least one woman was present, too, which evokes images of domestic family scenes more than a hunter's brief visit.

The pot's large size could reflect a need to feed large numbers of people, but its true function remains unknown for now. John proposed some alternative uses, as well. "Maybe it was used for storage between visits," he suggested. "Or maybe it was left there for other people to use out of good old-fashioned hospitality." According to historian Melissa Otis, a Tuscarora scholar once told her that Haudenosaunee people often left pots in rock shelters to inform newcomers that the place was already occupied.

Regardless of how they were used, clay vessels and their shattered remains have been found all over the Adirondacks. In his memoir, nineteenth-century hotelier James Wardner reported finding two large pots with flared tops and rounded bottoms near his inn at Rainbow Lake. One was exposed where a large birch tree had blown over, thus opening a hole where the roots had been, and the other was revealed by a fire that burned leaf litter away from the forest floor near Jones Pond.

An article that appeared in the *Plattsburgh Sentinel* in 1879 described how the Jones Pond pot was found: "The manner of its discovery was singular. He was following a deer over a fallow, from which, the fire had burned nearly all of the vegetable mold or peat. . . . Curiosity led him to stop and examine a hole made by one of the deer's feet, which revealed this earthen vessel, half buried in the sand, and into which the deer had stepped."

Wardner's collection of Indigenous ceramics and stone tools, like so many others from the region, has long since been lost to giveaways, theft, fire, and neglect. Nonetheless, pottery fragments have also been found along lakeshores elsewhere in the uplands. Such locations include Chateaugay Lake, Upper Saranac Lake, Lower Saint Regis Lake, Tupper Lake, Follensby Pond, Long Lake, Indian Lake, the Fulton Chain, Sacandaga Lake, and Piseco Lake. These discoveries help to demonstrate that the Adirondacks have been suitable for long-term occupation by Indigenous residents for centuries.

The earliest pottery found thus far in the Northeast dates back roughly 3,000 years and was apparently used for processing fish. With the spread of horticulture, pottery was mainly used for cooking domesticated plant foods that needed prolonged boiling. Indirect evidence for the clearing and burning of garden plots long ago might therefore be expected to appear in lake sediment deposits in the form of charcoal, silty runoff, or the pollen of crop plants.

Such evidence from sediment cores has been found at Crawford Lake, Ontario, where an Iroquoian farming settlement delivered charcoal, silt, and pollen from maize, squash, and sunflower fields into the adjacent lake from about AD 1300 to 1500. In hopes of finding our own version of the Crawford Lake story, some of my students and I recently collected sediment cores from Upper Saranac and Piseco Lakes where historical accounts reported Indigenous maize cultivation during the 1800s. Albert Fulton, a pollen expert at the University at Buffalo, examined subsamples of those cores but found no maize pollen in them.

Fulton explained that the results were not unexpected. "Maize pollen grains are relatively heavy," he told us, "so the wind won't normally carry them very far from a field."

The abundance of maize pollen at Crawford Lake was unusual, the chance result of wild geese pilfering meals from the fields and releasing their wastes in deep, oxygen-poor water where the conditions for preservation were ideal. Perhaps someday a sediment coring team will retrieve similar evidence of early horticulture in the Adirondacks from a small, deep lake that captured its own trove of ancient goose droppings.

Although we have not yet found crop pollen in Adirondack lake sediments, core samples from Wolf and Piseco Lakes do contain charcoal layers that represent forest fires of the distant past. Both cores registered more charcoal between AD 1000 and 1400 when unusually warm, dry climatic conditions were more conducive to fires. However, the abundance of charcoal at each lake peaked at different times, around AD 1100 at Wolf Lake and AD 1300 at Piseco. Were those separate fire peaks caused by local lightning strikes or were they set intentionally or otherwise by people? More studies of this kind might one day help solve that mystery.

Regardless of such technical details, the suitability of the uplands for intensive farming is beside the point when considering the human story of the Adirondacks as a whole. Traditional hunting, fishing, and gathering have sustained Adirondackers for much longer than the Three Sisters ever did.

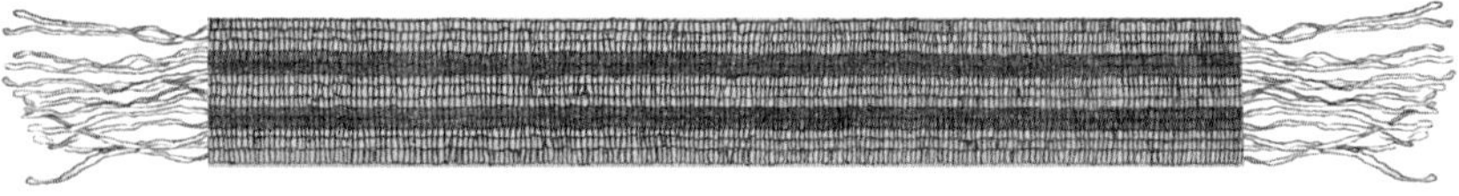

Let's not assume from the foregoing information that Indigenous horticulture was only about food crops. Another example of local plant domestication is tobacco, who is mentioned along with the Three Sisters in the Creation accounts of both Haudenosaunee and Wabanaki peoples. Tobacco accompanied or predated early food cropping in much of North America and eventually became one of the most widely used botanical products on Earth.

Nicotiana rusticum and *N. tabacum*, the two most common species, are thought to have been domesticated in the Andes Mountains between 8,000 and 6,000 years ago. When they later spread through Central America they were revered there as spiritual allies and agents of healing. Their leaves were often mixed with lime and pounded into a paste, which was carried in sacred gourds and consumed as oral snuff. Tobacco is related to deadly nightshade, and when the neurotoxic juices were absorbed in this manner they were powerful enough to induce visions.

By the time the first Europeans arrived, tobacco was being used for spiritual purposes throughout North America, mostly by smoking the dried leaves. Archaeologists who study the history of tobacco have developed ways of confirming this by extracting organic residues from the interiors of ancient smoking pipes of stone or clay to identify nicotine, the primary intoxicating agent in tobacco. One of the North Country's earliest nicotine finds came from a 2,300-year-old tubular pipe that was unearthed near Burlington, Vermont. To our knowledge no such chemical studies have yet been conducted in the Adirondacks, but an elbow-shaped pipe that was discovered at Sacandaga Lake shows that at least one person was smoking something, most likely tobacco, in the uplands several centuries ago.

The arrival of cultivated plants from Central America demonstrates that this region has long been a cultural crossroads, just as it is today. But as the next chapters will show, the influx of intruders from overseas during the last five centuries brought unprecedented disruptions to local ecosystems and societies. Many of the foreigners who helped create what we now call America did so through plagues, cultural erasure, warfare, slavery, and theft.

We must acknowledge the many painful aspects of that story and look back with honesty on the long, rough road behind us in order to see how far we've come and why continuing the journey toward a better world is worth the effort. It can also help us more fully understand and appreciate how the blending of diverse cultures has produced the complex tapestry of humanity that makes the Adirondacks what they are today.

THE GREAT DISRUPTION
(The Last 500 Years)

It is early summer on the knoll overlooking Lower Saint Regis Lake. The year is 1858, and we are watching a family who are going about their business beneath the pines on the sun-warmed crest of the slope. When not hunting and fishing here they spend much of their time downstream at Akwesasne where, a century earlier, their Abenaki forebears had been welcomed as war refugees by the Haudenosaunee community along the banks of Kaniatarowanenneh, also known as the Saint Lawrence River.

A small fire crackles and smokes at the center of a ring of stones where Mother and young Daughter are roasting a batch of plump white suckers. Father and young Son caught the fish earlier this morning in a nearby brook where lake-dwelling suckers gather to spawn at this time of year. Now they are repairing their net with dogbane-fiber cordage beside the fire while the catch is cooked.

A twig snaps, and a man's voice calls out from the wooded trail on the narrow neck of dry ground that links the knoll to the mainland.

"Hello-o-o, the knoll!"

The family members turn their heads toward the visitor, then to one another. Father rises and awaits the man's arrival.

"Good afternoon." A tall, well-built White fellow with a neatly trimmed beard extends his hand in greeting. "My name is Apollos Smith, but feel free to call me 'Paul' if you like." He appears to be in his thirties.

Father nods and replies, "Welcome to Akwesasne Lake. Will you join us for a meal?"

Smith takes a seat beside the fire and exchanges pleasantries with Father. Fish are passed around and savored, and tobacco pipes are lit. On the surface, this encounter seems enjoyable enough thus far, but the atmosphere changes when the purpose of the visit becomes clear.

"I have a difficult request to ask of you," Smith begins with trepidation now showing on his face. "Some friends of mine, including Mr. Wardner of Rainbow Lake, have suggested that I consider this location for a project I have in mind. 'About where the Indians are camping,' Wardner said, and this looks to me like the place."

Father glances at Mother, who rises and invites the kids to follow her to the lodges on the lee side of the knoll.

"I want to build a hotel here," Smith continues, turning to face Father squarely. "I hope to make it much grander than the one I've been operating at Loon Lake, and this section of shoreline is just the place for it. There's the Old Military Road nearby, the woods are full of game, and the lake is perfect for sport fishing. This knoll would make a nice picnic spot for my guests, too, with that spectacular view of the mountain."

He pauses to take a breath. "So I'm here today to make you and your companions an offer."

Father listens silently.

"I know you Indians have been used to camping here, but it simply can't continue when the hotel is built. Many of my guests wouldn't stand for it, and I don't think you would enjoy their presence either. I'd like to offer you another place to stay instead."

As polite as Smith is trying to be, it is clear that the offer is not really negotiable. Father senses this, too, but he lets Smith finish before responding.

"You can keep coming here but you'll have to camp farther away from the main hotel grounds. There are some nice little ponds just over there beside the road that I can make available to you instead of this spot. I know they aren't as easy to reach from the big lake, but nobody is likely to bother you there and you can use them for as long as you like."

Feel free now, Reader, to envision Father's response to this imaginary encounter, and those of the other residents of the knoll after Smith departed. We don't know exactly how the transition from Indigenous camping ground to hotel playground occurred, but we do know that Paul Smith's hotel opened in 1859 and that the knoll became known thereafter as "Picnic Point." In his memoir, Wardner mentioned a lakeside encampment of "Saint Regis Indians," presumably meaning residents of the Saint Regis community at Akwesasne, without specifying whether they were of Haudenosaunee or Wabanaki heritage. He also reported that the occupants were offered a chance to stay at what would later become known as Church Pond, but they chose to move elsewhere instead.

More than a century later, students at the college that now bears Paul Smith's name would know and love the beautiful place that Indigenous Adirondackers had known and loved for thousands of years before them. However, the college community was unaware of the deep history here until recently.

Ill-informed authors of the 1800s and 1900s had contributed to that lack of understanding with widely read narratives that excluded nearly everyone except their fellow Euro-Americans. In the fantasy worlds promoted by writers like Ralph Waldo Emerson, "Adirondack" Murray, and Alfred Donaldson, the uplands had been a howling wilderness devoid of people other than a few obscure primitives who only rarely visited the area. Being thus unoccupied, the reasoning went, the land was therefore conveniently ripe for exploitation.

To provide context for this phase of the Adirondack story, we must examine events dating back to the earliest arrival of Europeans in the region. Traditional watercraft also played an important but little-known role in the saga. We'll begin in more recent times, though, when the connections between the two families who inspired this book became intertwined with the story of an old boat that had once lain hidden on the bottom of Lake Placid.

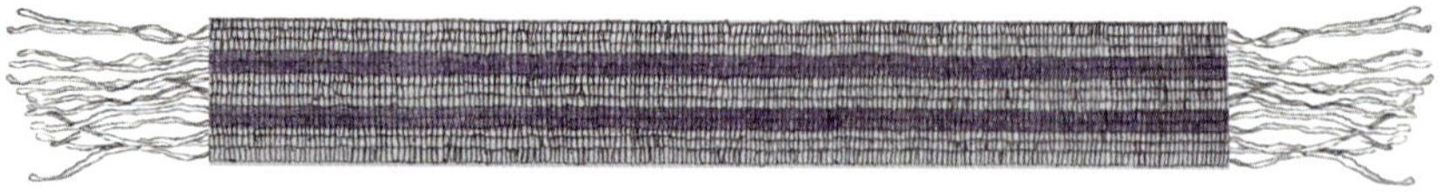

In the summer of 2014 my parents, Jay and Shirley Stager, celebrated their eightieth birthdays. Amid the feasting and reminiscing, I offered to take them to the Six Nations Indian Museum (now the Six Nations Iroquois Cultural Center), which was located just a few miles away from my residence in Paul Smiths. The place was always worth visiting, but there was also a more personal reason for this particular trip.

My parents had been involved in summer youth camping since the 1950s, particularly in camping with a social mission. When my mother was a teenager, she worked at a camp for children with developmental disabilities near Langhorne, Pennsylvania. Inspired by that experience, she later introduced her future husband to a career that would make him a prominent figure in the nation's youth camping movement, a feature of American culture that has borrowed heavily from Indigenous traditions.

By the early 1970s, he was running successful camps of his own in Maine that were based on what was then a pioneering philosophy of being coeducational, multicultural, and noncompetitive. That's when, unbeknownst to us, one of our earliest connections to the Faddens occurred.

As a teenager I shared traditional stories from the Northeast by firelight with children at Hidden Valley, one of our youth camps. In that manner I grew to adulthood amid Wabanaki tales of the spirit-hero Glooskap and the Haudenosaunee Creation account. One of my favorite go-to sources was a book by an author whose name I struggled to pronounce, and I kept the well-worn, smoke-scented copy in my office when I began teaching at Paul Smith's College in 1987. It was then that a student, Paul Hetzler, stopped by and noticed the book on my shelf.

"I see you have Ray Fadden's book of legends here," Paul said while reaching for it.

"No, that one's by a fellow named . . . um, Tehanetorens," I replied, mispronouncing it as usual.

Paul smiled. "Yes, I know. That's Ray. And his Mohawk name sounds more like 'deh-han-eh-DOH-lons' when you speak it."

It seemed that everyone in the North Country knew and admired Ray Fadden, even if they didn't know his traditional name. Many thought of him as the man whose long decades of teaching, writing, and social activism had helped inspire a grassroots revival of cultural pride at Akwesasne and elsewhere that still continues to grow. A museum with the shape of a Haudenosaunee longhouse that he and his wife Christine had established stood across the road from the house. Inside was a trove of hand-drawn charts and illustrations, wampum belts, the clay pot from the Silver Lake rock shelter, an old dugout canoe, and many other fascinating and informative relics.

Others simply knew Ray as the elderly fellow who fed wild black bears in his yard and who could—and gladly would—talk your ears off about Indigenous cultures, environmental issues, and American history. Like many before me I had delivered meat scraps to his door for the bears and had enjoyed visiting and listening to him at the museum. Now I was astonished to learn that, through his book, Ray had influenced me for much longer than I had realized.

I didn't find out until later that the connection ran even deeper. While researching this book, I learned from Saint Regis Mohawk tribal director Tsiorasa Cook-Barreiro that my long-standing interests in Indigenous cultures and history also trace back to Ray through the Akwesasne Mohawk Counselor Organization he launched during the early 1940s.

Ray's organization empowered young people to serve as proud cultural ambassadors to Boy Scout troops and summer youth camps across the country, and Tsiorasa's grandfather, William Cook, was one of those trainees. Cook informed and inspired several of the progressive Farm and Wilderness camps in Vermont, where he was known as "Wakio" or "Flying Cloud" to the staff and campers. He left them with a legacy of deep respect for Indigenous traditions before losing his life in a fighter jet training accident during the Korean War. Two decades later I attended his namesake camp, Flying Cloud, as a thirteen-year-old, after which I shared what I had learned there with youngsters at my own family's camp in Maine. I am humbled and honored to discover late in life that I, too, am part of the living legacies of Tehanetorens and Wakio.

All of this is to explain why my parents and I found ourselves shaking hands with Ray's son, John Kahionhes Fadden, at the cultural center several years after the great man's passing. I had expected us to enjoy retracing our common interests and family connections, and we did. But the encounter also launched a new phase of research into the story of the Adirondacks that led to this book.

That journey began when my father noticed the old dugout canoe on the floor and asked John how old it was.

"We don't know," John replied. "We haven't had the resources to date it."

My father turned to me and, in his usual blunt fashion, said, "You have a research grant. Why don't you have it dated for them?"

"Um . . . sure," I replied.

John turned to his son, David Kanietakeron Fadden, and said, "Why don't you help Curt sample the dugout when he's ready?"

"Um . . . sure," Dave replied.

The rest, both literally and figuratively, is history.

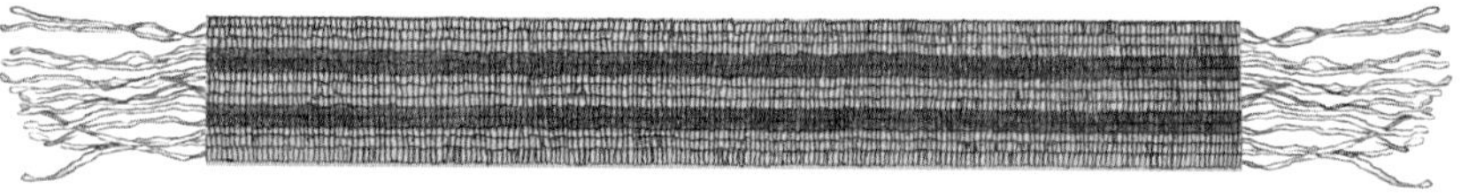

In 1960, a scuba diver named Jim Blackmore was searching the bottom of Lake Placid for old wood. During the heyday of Adirondack lumbering in the late nineteenth century, thousands of logs were cut, hauled to the water's edge, floated on the lake, and then flushed down the outlet stream for processing elsewhere. Some of those logs never made it to a sawmill because they sank instead. Salvage divers like Blackmore could sell the sunken relics for high prices because they were well preserved and had come from trees of great size and high quality. Such old-growth trees are now rare in the Northeast.

"I saw what I thought was an odd-looking log in about fifteen feet of water just off the Brewster Peninsula," Blackmore remembered when I spoke to him about it by phone. "After we hoisted it to the surface I looked over at the opposite shore and saw that famous singer Kate Smith waving to us from her property on the island there." Word of the find reached Ray Fadden, who bought the twelve-foot-long dugout canoe for twenty dollars and installed it in his museum.

Half a century later, Dave Fadden and I knelt beside the dugout at the request of our fathers. The cedar wood from which the boat was made was potentially suitable for radiocarbon dating, but the proper choice of samples was crucial for the technique to be most effective. One canoe that had been dredged from a wetland in Ohio in 1976 was a case in point.

The so-called Ringler dugout was carved from a twenty-foot-long section of a white oak tree. Initial analysis indicated that it was thousands of years old, but investigators later showed that it was actually made during or close to the early nineteenth century. Organic preservatives applied to the boat by museum staff had contaminated its surface layers with ancient carbon, making the dated wood sample seem much older than it actually was.

Preservatives were not a problem in the case of the Lake Placid dugout, which had been left untreated. Nevertheless, to avoid possible contamination from recent microbial growth or old organic mud that might have seeped into the wood while it lay on the lake bed, we used a tree-coring device to extract a slim cylinder of clean wood from the interior. Several weeks later Beta Analytic, a reputable dating lab in Florida, provided a much-anticipated answer to my father's query. The dugout was most likely built during the late 1600s to mid-1700s, but possibly even later.

Frankly, Dave and I were a bit disappointed with the results. We had hoped that this boat was more akin to a dinosaur raised from the dead like some ancient dugouts that have been recovered from lakes elsewhere. For instance, more than one hundred dugout canoes that were discovered during a drought-induced low stand at Newnan's Lake, Florida, ranged in age from a few centuries to 5,000 years old.

Our desire for the Lake Placid dugout to be ancient was not only based on aesthetics. It was also because a young age would make it more difficult to interpret for the museum. If the boat were more than 500 years old, then we could confidently attribute it to Indigenous Adirondackers on the basis of age alone. Both Indigenous and non-Indigenous peoples used similar watercraft here in recent centuries, making it harder to determine the origin of a boat from that time period. For instance, Winslow Homer's painting from 1870, *The Trapper*, features a Euro-American outdoorsman using a rustic dugout on an Adirondack lake. The Lake Placid boat might have been made by an Indigenous resident, but the young age and distinctive cut-marks left by sharp metal tools rather than blunter stone implements were also consistent with a Euro-American origin.

To investigate further, Dave and I compiled a list of other dugouts that had been reported from the uplands and resolved to study them all. Over the next several years most of our follow-up work in collaboration with anthropologist Christopher Wolff of SUNY Albany yielded similar results, and we eventually published our findings together in the *Adirondack Journal of Environmental Studies*.

Drill samples from a dugout discovered in Lake Ozonia that was on display at the Adirondack Museum in Blue Mountain Lake proved to be much younger than previous analyses had suggested. Beta Analytic's use of more reliable dating protocols, along with signs of metal tool use, placed its origin in or around the 1700s. A small one-person dugout, which was kept under the discoverer's house in Cranberry Lake, was so obviously marked by metal tools that it didn't require dating. Two others found in Utowana and Arbutus Lakes were also clearly modern. That process of elimination left us with only two candidates for oldest-known watercraft, and they were both owned by a resident of Wilmington who kept them out of public view in his home.

The dugouts were discovered in 1984 by a backhoe operator who was cutting a channel through a wetland at Twin Ponds in the northern uplands near Duane when his rig struggled to lift another load from the mire. Onlooker Jerry Bottcher later recalled for me the dredger's loud and unprintable reaction to seeing a twelve-foot-long, waterlogged monster emerge from the peat deposit. Soon after that, the second dugout was exhumed nearby. It was a whopping eighteen feet long.

Both watercraft were stored at the property owner's home in Wilmington despite requests from members of the Indigenous community to donate them to a public venue. The discovery and the controversy fired the imaginations of local residents and inspired Haudenosaunee poet Maurice Kenny to write about it.

"This is the story of four men who boarded a boat so many years ago," his poem read, in part. "Fog has covered footsteps, wind drowned voices."

Later in 2014, a mutual acquaintance arranged permission for Wolff and me to sample the Twin Ponds dugouts at the owner's home. The smaller boat, shiny with a coating of shellac, lay on the floor of an exercise room, and the larger one lay atop a bookshelf in the living room. It, too, had been coated with shellac, and one end was splattered with white paint. But the clean wood beneath the coatings on each boat was dark and blotchy with sooty patches, which indicated charring by fire. The surfaces were lumpy rather than planed smooth, and they lacked the neat linear slashes and notches that metal-edged tools had left on the other dugouts.

All of these features made it clear that the Twin Ponds watercraft were made using traditional methods. They had been laboriously charred, pounded, and chipped from tree trunks with fire and relatively blunt nonmetallic tools and then left on shore or submerged in the shallows for storage. The radiocarbon ages of wood cores, which we retrieved from the interiors beyond reach of the shellac coatings, showed that both dugouts were likely made during the 1600s, almost certainly by Indigenous Adirondackers.

Above left: The larger of the two Twin Ponds dugouts on a shelf in the owner's living room. *Above right:* Curt Stager and David Kanietakeron Fadden using an increment borer to extract a sample of wood from the Lake Placid dugout. PHOTOGRAPH BY JOHN KAHIONHES FADDEN

Opposite: The smaller of the Twin Ponds dugouts on the floor of the owner's gymnasium.

Below: Distinctive cut-marks on the wood of a dugout that reflect the use of sharp metal tools.

The discoveries at Twin Ponds provide compelling evidence that Indigenous people of the past spent significant amounts of time in the uplands. One isn't likely to interrupt a quick trip over the mountains to spend days or weeks constructing a massive dugout and then abandon it. Moreover, the boats were much too large to navigate the shallow outlet stream, indicating that they were made to be used right there on Twin Ponds. Their low and uneven profiles would also have made the interiors too wet for hauling perishable gear over long distances, so they were most likely used as platforms for fishing, hunting, trapping, and perhaps even for recreation.

Why this focus on dugouts rather than other kinds of canoe? It's because dugouts are the oldest known boats in the region. We have no firm evidence from past millennia of wood-framed canoes covered in bark such as those that were commonly used by Indigenous Adirondackers in recent centuries. They decay quickly and are therefore easily lost to the ages, and it is even possible that they weren't in wide use in these parts until recently. The oldest known birchbark canoe still in existence is one from Canada that was brought to England by a British officer shortly after the American Revolution.

And how do these boats and their makers fit into the story of life in the Adirondacks? To answer that question, we must first consider events that followed the arrival of Europeans in the North Country. The account that follows is based on information provided by Darren Bonaparte, director of the Saint Regis Mohawk Tribal Historic Preservation Office, as well as the writings of Cindy Amrhein, Stephen Sulavik, Melissa Otis, and other scholars.

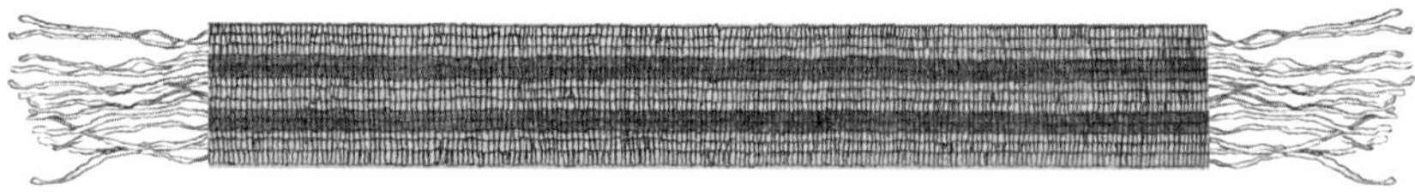

If we disregard a possible brief visit from Norse seafolk roughly 1,000 years ago, some of the first recorded encounters between foreign intruders and peoples of the Adirondack region occurred during the 1500s, when European fishermen and explorers sailed up the Saint Lawrence River. By then, the cultures and languages of the region's Iroquoian and Algonquian nations were close to their present forms, and a complex web of interactions had developed among residents of the North Country including the Haudenosaunee (primarily the Kanienkehaka), Wabanaki (primarily the Western Abenaki), and others.

Many of the specifics of who lived where and for how long remain open to debate. Oral traditions and most academic scholars hold that Haudenosaunee ancestors had moved into the region from the Ohio Valley or Appalachian Mountains relatively recently. Some place that arrival just a few centuries prior to European contact while others place it a millennium or more earlier, but all agree that the Haudenosaunee were not the North Country's first inhabitants. Bonaparte has described a Kanienkehaka oral account from the early 1700s as noting that the Haudenosaunee of the Saint Lawrence Valley "are not unaware that they are foreigners in the lands in which they inhabit at present." According to many sources, the previous inhabitants of the Adirondacks more likely gave rise to Algonquian peoples who were largely displaced from the uplands sometime after the Haudenosaunee arrived.

Exactly how did such territorial changes occur? Much of that tale has been lost to history, but the presence of fortified villages in the Saint Lawrence watershed during the last millennium hints at violence due to frictions among Haudenosaunee, Wabanaki, Wendat, and other Indigenous peoples of the region. The first European arrivals therefore encountered a dynamic mosaic of local cultures, conflicts, and alliances with shifting territories and trade networks, and the foreigners soon added new layers of complexity to that mosaic.

During the 1500s and 1600s, the availability of foreign trade items and firearms fueled a search for more animal pelts to pay for them. This commerce further intensified territorial rivalries, alternately forging and destabilizing relationships among various European and Indigenous peoples. The Five Nations confederation of Haudenosaunee peoples is widely thought to have formed in what is now Central New York during or close to that period, later to become the Six Nations when the Tuscarora Nation joined the alliance. The Wabanaki confederation of Abenaki, Penobscot, and other nations formed in northern New England around that time, as well, perhaps in response to conflicts with the Haudenosaunee and Europeans.

It was a period of great instability in the North Country, details of which are better documented in the lowlands than in the Adirondack uplands where foreign observers rarely ventured. In 1535, when French explorer Jacques Cartier visited large Indigenous settlements along the Saint Lawrence River, his journals described a thriving society that academics have labeled the "Saint Lawrence Iroquoians." By the early 1600s, many residents of those settlements had dispersed for reasons and destinations that remain as yet unclear.

The swelling European presence amplified long-standing tensions among Algonquian and Iroquoian residents of the North Country. Abenaki peoples who lived near Lake Champlain and knew it as Pitawbakw ("Twin Lake") are thought to have included parts of the eastern Adirondacks in their territories during the last several centuries if not earlier. They came into more conflict with the Kanienkehaka, who occupied much of the uplands then. Samuel de Champlain further inflamed those tensions in 1609 when he participated in a skirmish between his Algonquian companions and their Haudenosaunee opponents on the shore of the lake that now bears his name. Such foreign interferences in the region helped strengthen alliances among certain Algonquians and the French, who later faced off against certain Haudenosaunee and the British.

Religious missionaries also arrived during the 1600s, and the eyewitness account of one Jesuit missionary, Father Isaac Jogues, refutes claims that Indigenous people avoided the Adirondacks in winter because the conditions were too cold. During his captivity among the Kanienkehaka in the winter of 1642, Jogues endured a frigid trek through the snow to an upland location that Sulavik argues was probably in the Saranac

Lake region. Along with the men, who spent most of their time hunting deer in the wintery forest, were women, children, and elders as well as the hapless Jogues.

By the mid-1600s, intertribal conflicts over territories and trade were displacing Indigenous communities throughout the Northeast, and overhunting threatened to extirpate the beavers whose pelts were being sold to European fur merchants. The Twin Ponds dugouts were built during that time frame, presumably by people who were using the uplands for hunting, fishing, or commercial trapping. We may never know what nation the makers of those dugouts belonged to, but one reasonable possibility is the Kanienkehaka, who were much involved in the upland fur trade then.

Even more devastating than the warfare were multiple plagues of smallpox, measles, and other imported diseases that killed Indigenous people throughout the Northeast on a scale rivaling that of the Black Death in Europe. Historical estimates of death tolls at sites across the region range as high as 90 percent or more. Small wonder, then, that early accounts by foreign visitors reported that people were scarce in the Adirondacks. Traumatized survivors of violence and pestilence during the terrible 1600s found refuge where they could, stood their ground when possible, and did whatever it took to keep themselves and their cultures alive.

During the 1700s, conflicts and disease continued as the trickle of foreign settlers became a torrent and international tensions intensified. By the 1750s another coalition, the Seven Nations of Canada, had united Wendat, Abenaki, and northern Kanienkehaka communities along the Saint Lawrence River and was comparable in power to the Six Nations league farther south. Meanwhile, the region's web of Indigenous cultures was being drawn into escalating conflicts between French and British interests that culminated in the French and Indian War.

In 1755, Haudenosaunee fighters who were allied with French forces found themselves facing combat with kinsmen who supported the British at the Battle of Lake George. For years, reciprocal raids between the Haudenosaunee and Wabanaki also spread fear and suffering as far east as Maine and might have discouraged Abenaki people from using the Adirondack uplands for a time. And yet, when Robert Rogers and his rangers brutally assaulted the Abenaki settlement of Odanak near Montreal in 1759, other Indigenous communities including the Haudenosaunee of Akwesasne sheltered many of the refugees, some of whom remained with them permanently.

The Adirondack uplands might have been somewhat protected from such mayhem because most of the Europeans had settled in the lowlands where farming and commerce were easier for them. Melissa Otis suggests that the uplands thereby became a "place of refuge" where Indigenous people were less subject to foreign interference. In that situation, it would not be surprising if outsiders were told that the Adirondacks were empty and not worth visiting.

All in all, the deadly effects of plagues and seemingly endless warfare made the 1600s and 1700s a poor time frame upon which to base our impressions of upland population numbers through the ages, not to mention the lack of comprehensive census records back then.

The Catholic mission village of Saint Regis was founded around 1755 beside the river at Akwesasne. By that time, many of the region's Haudenosaunee residents had become Catholic, but most still maintained traditional societal structures such as clans, matrilineal families, and a political system in which chiefs were established by clan mothers. Many Kanienkehaka also left their homes in Central New York to avoid encroachments by Euro-American settlers and to find sanctuary at Akwesasne.

Later in the 1700s, shifting international relationships brought further difficulties to the Haudenosaunee, many of whom had established close ties with the British during or after the French and Indian War and therefore did so again during the War of Independence. When the American Revolution finally drove Britain from the colonies, many of the victors tended to treat Indigenous residents as obstacles or former enemy combatants regardless of whether or not they had been British allies. It was during that tumultuous period that most of the Adirondack uplands ended up in Euro-American hands.

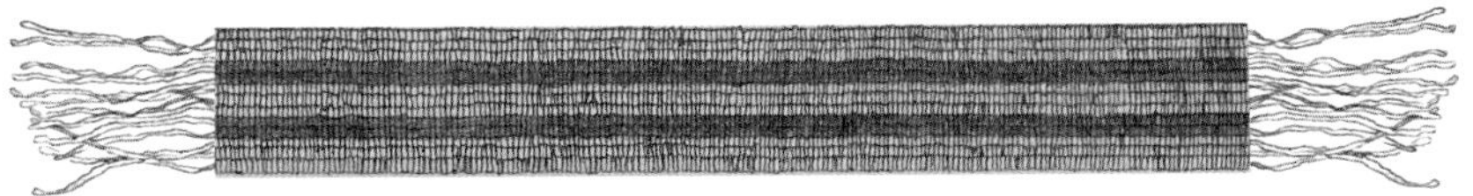

It has recently become common for public events across the continent to begin with a formal land acknowledgment that reminds listeners of local history. In the Adirondacks, the statement often notes that "we are standing on the traditional homeland of Indigenous peoples," but few who use those words are aware of exactly how, when, or from whom any particular piece of this vast territory was transferred. We will now address the basics of that story here.

The process involved a convoluted series of events that unfolded during the 1700s among foreign nations and Indigenous occupants of the uplands, particularly the Haudenosaunee. It included both legal and illegal financial transactions, differing cultural concepts of governance and landownership, and age-old rights of conquest. Widely respected historians including Bonaparte, Amrhein, Sulavik, and Otis have investigated the saga in depth, and we summarize key aspects of their findings here.

Most of the Adirondack uplands were transferred in three enormous parcels around the time of the American Revolution. The first big transfer originated in 1772, when a group of investors worked with representatives of the British Crown to buy more than a million acres in the central highlands from chiefs representing Kanienkehaka communities of the Saint Lawrence Valley. Agents for that "Totten and Crossfield Purchase" paid the chiefs for deeded title to the land in a legal process that, according to Darren Bonaparte, apparently faced no documented objections from Haudenosaunee leadership at the time. However, the final transfer of property was not yet fully completed when the American Revolution derailed it. The loyalist investors fled to Canada and the land was seized by the state of New York, after which it was sold off piecemeal to various buyers. The huge triangular tract covered most of what is today's Hamilton County, extending eastward through Raquette Lake into the High

Peaks region of Essex County and thence southward into Warren County around Lake Pleasant and the Fulton Chain of lakes.

The groundwork for the other two mega-transfers was also linked to the American Revolution. In 1783, the Treaty of Paris ended the war and saw Britain cede lands in the former colony of New York to the new United States without significant input from Indigenous residents. The treaty split the North Country into separate states and nations, which isolated communities and families who had previously traveled freely in the region. Treating large areas of Upstate New York more or less as spoils of war, state officials declared those areas "unoccupied" or "abandoned" by their Indigenous inhabitants and began to offer parcels for sale or as compensation to military personnel in lieu of cash payments. Whether the state, in some cases acting independently of the federal government in violation of a standing "Nonintercourse Act," was legally empowered to do all of this remains open to question. Regardless, we see in that process a practical consequence of the myth of absence that persists to this day.

The second great transfer of Adirondack lands occurred in 1786 amid that post-Revolution turmoil, when representatives of the state began to offer more than 650,000 acres of the northeastern Adirondacks to military personnel. Known today as the "Old Military Tract" that included what are now the towns of North Elba, Saint Armand, Franklin, and Bellmont, the offer had few takers. The rugged high country was less desirable to most Americans at the time than more developed sectors of Central New York where a successful "New Military Tract" giveaway took place several years later. Many of the upland parcels were eventually sold off bit by bit or remained state property, and the tract's story is now so commonly forgotten that an eponymous Old Military Road that passes through it leads many users to believe that the route's name simply reflects usage by soldiers in the wars of centuries past.

The third and largest transfer took place during the early 1790s when the new state of New York offered the gigantic "Macomb's Purchase" to a land speculator. The roughly four-million-acre tract encompassed the northern uplands from the Tri-Lakes area, including the knoll on Lower Saint Regis Lake, through the northwestern corner of today's Adirondack State Park to Lake Ontario.

Lingering disputes with the state over the sale or seizure of lands led to additional payments to representatives of the Kanienkehaka and Seven Nations of Canada, transactions that have themselves remained controversial to this day. In simplest terms, however, most of today's state-owned wilderness in the Adirondacks stems from the three largest land transfers of the 1700s.

Macomb's Purchase also set aside six square miles along the Saint Lawrence River for the Saint Regis Mohawk community at Akwesasne. Today, the Saint Regis Mohawk Reservation on the New York shoreline is linked by treaty access rights to the Akwesasne Reserve on the Canadian shore. A coalition of traditional chiefs of Akwesasne and elected tribal leaders from both sides of the river governs the community with only minimal jurisdiction from the governments of Canada and the United States.

It is important to remember here that the extensively documented territorial interactions among Euro-Americans and the Kanienkehaka, which we too have emphasized on these pages, tend to overshadow historical connections of other Indigenous peoples to the uplands. Oneidas also considered some of the western sectors of the uplands to be part of their territory. The Kanienkehaka are said to have driven Mahican people from the southern Adirondacks during the 1600s. They might also have done the same to the Wendat in what is now the Ogdensburg-Oswegatchie area as well as to Wabanaki peoples of the Champlain watershed. Nonetheless, some Indigenous people other than the Haudenosaunee were still residing in the high country during the nineteenth century, if not earlier.

According to Melissa Otis, Piseco Lake was named after a Haudenosaunee resident of its shores, but most accounts of Indigenous Adirondackers from the 1800s refer to people of Wabanaki heritage. Adirondack guide Carlos Whitney reported seeing an Indigenous woman ice-fishing on Upper Saranac Lake, where a community of likely Abenaki descent raised summer crops at the so-called Indian Carry. The memoir of James Wardner mentioned Abenaki people who traveled widely in the uplands by canoe and wintered in the Lake Luzerne area. The noted guide and hunter Moses "Sangemo" Saint-Germain, has a pond named after him and was, according to one of his descendants, of Wabanaki and French heritage. Other Wabanaki residents who became widely known as guides, hoteliers, hunters, or trappers included members of the Sabattis and Benedict families.

It is a complicated legacy, but despite the many difficult trials of the last five centuries Indigenous peoples maintained their long-standing presence in the Adirondacks through it all. They adapted to current circumstances as their ancestors did in previous millennia, and they brought their cultures successfully into the present day. However, centuries of interaction among peoples of diverse backgrounds have made it difficult in some cases to distinguish aspects of modern Adirondack culture that are of strictly Indigenous origin from those with foreign roots. In particular, the names of local landforms are sometimes thought to be traditional when, in fact, they are not. Let's now examine some of those labels, beginning with the name of the region itself.

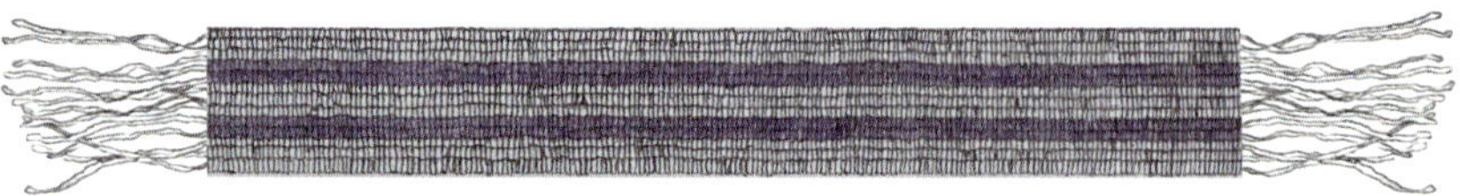

The mountains we now call the Adirondacks were not given that name by Indigenous residents of the North Country, although the term "Adirondack" is derived from a Haudenosaunee word. It can be translated as "porcupine," "they eat trees," or "bark-eater," and various authors have proposed differing versions of the root word including Atirutaks, Anentaks, or Ratirontacks. None of them, however, are pronounced "add-ih-RON-daks." The last term, for instance, is pronounced more like "lah-dee-LOON-duks," which suggests that today's label should more properly sound closer to "ah-dee-LOON-duks." Regardless of its spelling or pronunciation, however, the source term was originally used in reference to people, not the mountains themselves.

According to sources including Frank Speck, Dyneley Prince, Russell Carson, and Stephen Sulavik, Kanienkehaka people used some form of the word in recent centuries as a pejorative aimed at Algonquian peoples in general, particularly those such as the Innu people (also known as "Montagnais") who lived on or near the northern side of the Saint Lawrence Valley.

The label has since been interpreted in several ways. According to Prince, it might have referred to an Algonquian habit of clearing fallen trees from rivers to enable canoes to pass. Doug George-Kanentiio, a former editor of *Akwesasne Notes*, has proposed that the name referred to the practice of flavoring food with powdered bark. Other sources claim that the term referred to the use of the inner bark (cambium) of trees as starvation food, something that all peoples of the region are likely to have resorted to on occasion.

Having been introduced to "pine dining" by one of my students, Alex Comer, I can attest that cambium is indeed edible when thin strips of it are roasted until crunchy or ground into powder to thicken a stew. Pine cambium contains sap and living tissues that offer some nutrition in a pinch, but not much. For Haudenosaunee who lived in large agricultural settlements, a "bark-eater" nickname could have implied that their Algonquian rivals were incompetent or old-fashioned because of their greater reliance on traditional hunting-gathering lifestyles. In that sense, the insult brings to mind city folk who view rural folk with scorn.

Regardless of the derivation, the label was not originally applied to the mountains. Nonetheless, some local folklore has morphed it into a pseudo-traditional place-name that can also help perpetuate the idea of the uplands being uninhabitable.

A self-published novel by a local author contains an example of this in which one of the characters explains that "Adirondack" translates to "Land of the Bark Eaters." The term was supposedly meant to "paint a picture of a wilderness so merciless, so brutal and unforgiving, that if you ever get lost in it you will starve and be reduced to peeling the bark off trees and eating it."

Again, let us remember that the "Adirondack" term applied to people, not peaks, and it arose from interactions between Haudenosaunee and Algonquian nations in and around the Saint Lawrence Valley. The insults flew both ways, of course. Wabanaki peoples are said to have called the Kanienkehaka "Magowak," which later morphed into "Mohawk." Some sources translate it into fearsome terms such as "man-eaters" or "cannibals," but when I recently asked Abenaki scholar Paul-René Tamburro about it, he offered a different take on the matter. His translation is more akin to "coward" and is not necessarily about eating at all. In essence, he explained with a wry smile, it can refer to the opposite bodily function that may occur to the rear when one is fleeing in terror.

The lack of clarity on the deeper origins of the term "Adirondack" is irrelevant, however, when it comes to the landscape. It was first formally applied to the High Peaks by Ebenezer Emmons, a Euro-American who surveyed them during the 1830s, and it subsequently encompassed the rest of the uplands. He chose the name to commemorate "a well known tribe of Indians who once hunted here," as he explained in a state document published in 1838. The tribe he had in mind might presumably have been one of the aforementioned Algonquian peoples of the Saint Lawrence Valley although the name, quite naturally, was not one that they seem likely to have used for themselves.

Emmons also named the tallest peak "Mount Marcy" after then-current governor of New York, William L. Marcy. Around that time, the poet and author Charles Fenno Hoffman informally applied the name "Tahawus" ("cloud-splitter") to the same peak. According to author Russell Carson, Hoffman derived it from a Seneca term that translates to "he splits the sky."

Emmons and Hoffman surely meant well and wished to honor Indigenous residents of the North Country by naming the mountains as they did. However, a more respectful way to honor those residents might have been to continue using names by which the landforms were already known. In any case, the labels these men chose are not generally pronounced according to Haudenosaunee speech patterns (see the glossary), and the Adirondack name, though easy on the ears and rooted in a traditional local language, can seem rather odd upon closer examination.

Imagine, if you will, that Germany had won World War II and occupied the United States. Imagine that those conquerors renamed the island that New York City now sprawls upon with a term that a German mapmaker overheard when local city-dwellers disparaged folks from a distant, rural part of the state. The term was insulting and had nothing to do with the island, but he liked the sound of it so he "commemorated local history" by adding the label to official German charts. In that hypothetical tale, Manhattan thereby became known as "Hillbilly Island" and its urban inhabitants were called "hillbillies." Add a Germanic twist to the pronunciation as well, perhaps "Hill-Wilhelms," and it's easy to see why it wouldn't be the most flattering—or accurate—tribute.

Well then, what did local residents really call this place in ages past? We'll never know what labels the earliest Adirondackers used, but we do know some of the names from more recent centuries. Melissa Otis reports that various Kanienkeha labels for the Adirondacks could be translated as "The Mountains" or "Big Mountains." Some residents of Akwesasne have suggested that the peaks were named in reference to milk or breasts, in much the same fashion that a French word was applied to the conical Teton peaks in Wyoming. Other residents report that Kanienkehaka people traditionally described the uplands with various terms that can translate to "The Place of Snow" (Kohserake) or "The Place of Animals" (Tsikariotanakere).

Wabanaki peoples of the region also used landform names of their own. In an interview with Dyneley Prince at the close of the 1800s, Abenaki wilderness guide Mitchell Sabattis reported that the High Peaks were known as Wawonbadenek ("White Mountains") in reference to their snowy mantles.

Some local landforms still carry their previous labels, or at least anglicized versions of them. According to Sabattis and other sources, many descriptive lake names were much the same when translated between English and Abenaki. The current name of Long Lake, for instance, is a close translation of Kwenogamak ("At The Long Lake"), and the same is true for Lake Clear and Blue Mountain Lake. On the other hand, Tupper Lake was formerly known as Side Lake or Branch Lake (Paskangamak, in Abenaki) because of its position astride what is now called the Raquette River. The Saranac Lakes derive their present names from an Abenaki word that translates to "sumac," and the Saint Lawrence River was Kchitegw, or "Great River." Lower Saint Regis Lake, where our time-traveling knoll is located, was known to Abenaki residents as Pokuizasne Nepes, which translates to "Akwesasne Lake." That name refers to the community of Akwesasne at the mouth of the outlet river, presumably because this lake was the first one a traveler from that community would encounter on an upstream journey into the high country.

The experiences of Indigenous Adirondackers during last few centuries have been both painful and inspiring, and the often-turbulent confluence of cultures in these uplands has carried many elements of the past into the present, as well. Although interpretations of that history may vary, one thing is certain: Indigenous people lived in and around the Adirondacks for thousands of years, and their descendants are still living here in modern times.

In recent decades a cultural renaissance has blossomed among Indigenous residents of the North Country. Its wonderful manifestations are too numerous to list in total here, but they include the Abenaki Heritage Weekend, the Kahnawake Pow-Wow, the Akwesasne Cultural Center and Six Nations Iroquois Cultural Center, numerous written works by and about Wabanaki and Haudenosaunee people, the Akwesasne Freedom

School and Native North American Traveling College, the pottery and documentary films of Katsitsionni Fox, the loom beadwork of Kelly and Tyson Back, the music of Theresa Bear Fox, Joanne Shenandoah, and others, and even a new Marvel super-heroine whose character was heavily influenced by members of the Akwesasne community.

Dave Kanietakeron Fadden will now introduce some of his personal insights into this and other, more troubling aspects of recent Adirondack history in our closing chapter.

KANIETAKERON
2016

TWO ROWS, TWO WORLDS

I often hear folks say that Native peoples didn't live in these mountains because it is too harshly cold here in winter and that they only hunted during warmer months, returning to the valleys where the main villages were. I never believed this notion. I live in the tiny Adirondack hamlet of Onchiota, and I also live at Akwesasne along the Saint Lawrence River. It's cold down at Akwesasne, too, sometimes colder than Onchiota.

Our ancestors were tough and had skills to survive in any environment. They knew what to eat and how to get it. They had the necessary clothing. They had fire. They had homes. Native people live in the harshly cold Arctic comfortably and have done so for many generations. There is a beauty in the mountains, the place that my ancestors called Tsikariotanakere and that is beloved by many people today. Why wouldn't Indigenous people have loved the Adirondacks in the past, as well?

I suspect that some aspects of the past are reflected in the habits of many people today. During certain seasons people travel to the mountains to hike, ski, camp, or simply enjoy the fresh air and clean water. Then many of them return to the towns and villages they came from while other folks remain here year-round. I imagine that it's similar to what happened among many Indigenous peoples in centuries past. As far as we know there were no large villages in the uplands, but rather small communities of several families who loved Tsikariotanakere and were well prepared for the harsh winters.

Our Six Nations Iroquois Cultural Center collections include many pre-Columbian artifacts from the region, most of which were found by accident. Until recently, there have been very few archaeological studies in these mountains. Several years ago, my father attended a conference and listened to an archaeology professor from the University of Vermont who showed a map indicating areas where Native artifacts were found. Dad noticed that on the Vermont side of the map there were sites dotted all over the Green Mountains and along the shoreline of Lake Champlain. There were, however, very few in the Adirondack uplands. My dad raised his hand and asked why there were so few sites in the Adirondacks, seeing as the Green Mountains and Adirondacks are very similar. The professor simply stated that it was because he and his colleagues studied places that are close to their homes in Vermont, and that they lacked the resources to study the Adirondacks in depth, as well.

There was a concept behind numerous agreements and treaties that were made between the Haudenosaunee and Euro-Americans as well as other Indigenous nations that can serve as a theme for my words here. The Two Row Wampum is a beaded belt whose design consists of two parallel rows of dark purple wampum beads surrounded by white wampum. There are various depictions of belts with a similar design that represent a pledge of peace and friendship between two peoples.

One row of dark wampum represents the Haudenosaunee in our canoe paddling down the river of life. Our beliefs, customs, and laws are contained within this canoe. The other row represents the other party in this solemn agreement, the Dutch, English, French, and later the Americans sailing in their winged ship down the same river of life. Safely tucked away in our respective ships are our beliefs, customs, and laws. These two vessels travel together and never interfere with one another. Between them the white wampum represents peace, friendship, and mutual aid. I remember learning about this concept at a very early age, and I still live by it. We are different but we are friends. We are to respect one another's ways so long as they do not interfere with our individual paths down the river.

Personally, I still hold to that agreement and have traveled the waters of my own life in recognition of the two rows. I have lived among those in the ship as well as the canoe. I have learned that there are great differences between the two paths but also much in common, something that I have also found to be true with members of cultures from around the globe who have visited our museum in Onchiota. Our exchanges of thoughts and philosophies are remarkably similar in certain aspects and yet very different in others. My grandparents and parents taught me to respect all peoples no matter their beliefs as long as they do not try to change the direction of our own canoe.

I was very fortunate to grow up next door to my grandfather, Ray Tehanetorens Fadden, my grandmother Christine, and the modest museum in Onchiota that they built with their own hands in 1954. They called it the Six Nations Indian Museum and filled it with an impressive collection of Native artifacts, handcrafted informational charts, pictographic beaded belts, and elaborately painted floors. They welcomed visitors from Native communities, local Adirondack towns, and other places around the world. Originally, the museum opened during the warmer months and closed after Labor Day when my grandparents returned to Akwesasne so my grandfather could teach at the Mohawk School in Hogansburg. My parents, John and Eva Fadden, took charge of the museum during the 1980s and ran it until very recently, when I took the reins myself as the third generation of our family to do so.

When I was old enough, I would watch and listen as my grandparents interacted with hundreds of visitors from all walks of life. One of my jobs was to watch for anyone who wanted to touch the artifacts and, perhaps, slip one of them into a pocket or bag. Being rather shy, I sat quietly on a fur-covered bed in the farthest room and tried to be invisible. People more or less ignored me and that was fine by me. I just prayed that they wouldn't try to steal anything, because I would then have to raise the alarm and draw attention to myself.

Through all of those years at the museum, I was also a student as well as a lookout. My grandfather was a world-class storyteller and lecturer, and his words permeated my young mind. I loved the legends and stories he told to visitors. I would imagine the giant bear, taller than the trees, who terrorized the Mohawks in the village, ate all of the food in the area, and ultimately found a trail into the sky to become the Big Dipper constellation. I pictured the Thunder Being taking a beautiful Mohawk woman into the clouds and marrying her. I learned that I was walking on the back of a giant turtle, and whenever I felt an earthquake I knew that it was the turtle stretching his back. My grandfather's words molded my creative spirit and eventually led me to become an artist and illustrator as an adult.

Grandpa's delivery of the legends was captivating, and he would also include tidbits of historical and cultural information in them. His tone and cadence were crafted in such a way that the audience would often sit with eyes wide and mouths agape. However, when he delivered his lectures about history, his words shot out like bullets from a machine gun, rapid-fire and powerful with information and facts. It was hard to keep up at times. He wanted to inform his audience as much as possible before they left the museum, especially if they were teachers. The history he related was often difficult to listen to, as the treatment of Indigenous people he described was brutal, terrifying, and grotesque. He didn't sugarcoat the facts, and I remember seeing tears run down the faces of some listeners as they learned about the horrors that so many Native people have endured.

As a teacher during the decades following World War II, my grandfather saw that the American educational system overlooked the gifts Native peoples have given to the world in terms of food cultivation, technology, and thought. The school system was highly Eurocentric when presenting the cultures and societies of Indigenous nations and people of color in general. He had to fight the school administration to teach Mohawk children the true history of this country and instill a sense of pride in their own culture and worth.

One time my grandfather had to shut the window blinds and have a student stand in the hallway on the lookout for administrators in order for him to teach the class about some of their own histories and heroes. He saw a systematic degradation of Indigenous culture throughout the school system as well as in the American and Canadian policies of forced assimilation through mandatory attendance at Indian residential schools.

The cultural genocide that was thereby inflicted upon Indigenous children has left a dark stain on our history. From the late nineteenth century to the 1990s, the governments of Canada and the United States funded church-run schools in an attempt to eliminate all aspects of the culture and spiritual beliefs of First Nations children. Richard Pratt, superintendent of the Carlisle Indian Industrial School in Pennsylvania, summarized the policy in a simple, brutal phrase: "Kill the Indian, save the man."

Thousands of Native children were coerced, tricked, or forced to journey to these schools, which were usually located hundreds of miles from home, with the promise of plentiful food, education, and housing. Upon their arrival, however, anything remotely reflecting their culture was tossed away. The children had their hair cut short. Their traditional clothes were discarded and replaced by military-style uniforms. When they spoke in the only language they knew, they were punished, sometimes severely.

The harsh settings led to physical, sexual, and mental abuse that caused lasting, multigenerational harm. Those who survived the schools (many died) later went home to find themselves with no sense of identity, no language, and no culture. In turn, they often failed to teach their own children their traditional language or culture because they had had it stripped from themselves as children. This, of course, was the intended outcome of the policies.

One elder related her experience as a four-year-old girl at the Mohawk Institute Residential School in Ontario, which was known as "The Mush Hole" because of its terrible food.

> *They'd throw us in this dark room. They'd tell us the rats were gonna get us. I didn't then know why I was being thrown in there and I used to wonder what did I do, and I would cry. And Rosemary would cry. We cried and cried for hours in there, not knowing why we were*

> *in there. And when I learned a little bit of English I knew they were throwing us in there because we didn't speak English. I must have been stubborn because [I thought, if] you want me to speak English, I won't speak English. So, I didn't speak at all for two whole years.*

This anecdote is one of thousands of experiences that Native children as young as four years old endured. At the policy's peak, there were 130 Indian boarding schools in Canada alone, with the last one closing in 1996. Another 408 schools operated in the United States.

Most American citizens are unaware of this dark chapter of history, and such ignorance has contributed to misinformed and cruelly racist views of Indigenous peoples. My grandfather, Tehanetorens, made it his

Above: One of the rooms inside the original Six Nations Iroquois Cultural Center.

Opposite: David Kanietakeron Fadden helping to keep oral traditions of the Haudenosaunee alive at the cultural center.

life's work to change such attitudes and to instill pride and a sense of identity in the minds of his Mohawk students. His lifelong mission has become a major part of my own life, too.

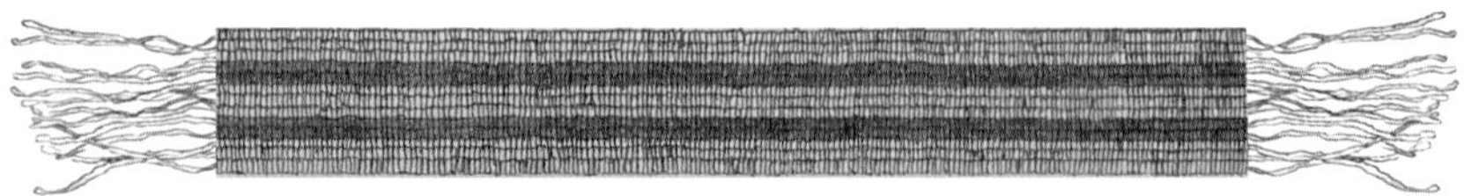

My experience as a student in the Saranac Lake School District was largely positive, but I encountered anti-Indigenous racism for the first time when I was in the third grade. While wandering around the playground during lunch recess, I overheard a conversation about the school district's sports team, the Saranac Lake Redskins. I was aware of the Redskins name but had never given it much thought. At the time, I was more concerned about trading baseball cards and playing with *Star Wars* action figures. The Redskins logo was of a Native man in profile with a pair of dangling yellow feathers. Many of my relatives and friends at Akwesasne resembled that man, and I liked the image as a result. But that feeling changed during third grade.

On that day a classmate pointed to me and asked loudly, "Aren't you a redskin, Dave?" Then came more racist comments from my so-called friends.

"Hang on to your scalp!"

"Where's your bow and arrow?"

"Do you live in a tipi?"

In that instant, I realized that I was different from my classmates. Although some of them had asked me about my culture in a respectful manner and some didn't know or care that I was a Mohawk, it felt like the cruel words from the others were directed at my family, my culture, and my heroes. I was extremely proud to go to the longhouse to participate in traditional ceremonies. I was proud to learn of the Peacemaker, Ayenwatha, Pontiac, Tecumseh, Sequoia, Jigonsaseh, and others. I was proud to speak the Mohawk language with my family. My classmates of European descent had no clue that what happened on that day affected me deeply, and during my remaining years at school I "hid in the back row" as I had done as a lookout in the museum.

All Indigenous people have heard similar racist comments, and still endure them when, say, the Kansas City Chiefs play a football game. The tomahawk chop is alive and well in Atlanta when the Braves play. Countless school districts fight to keep their redskin, Indian, or warrior mascot names in place, insisting that they are honoring Native Americans in doing so. This disregard for another culture's identity is not limited to sports enthusiasts; it is everywhere.

At the museum we have to maintain a thick skin and realize that most folks utter such words out of ignorance without realizing how harmful they are, although sometimes the words really are meant to hurt or insult. Regardless of the intent, our job at the museum has been to challenge untruths and insensitive comments about Indigenous people with carefully chosen words of our own, and I prepare myself to do so every morning before we open.

As an adult I have learned to control my emotions when confronted with hurtful language, and I do my best to educate and enlighten rather than respond in anger. There are times, however, when it comes out of the blue from the last place or person I expect it from.

During my college years I worked as an intern for the Department of Environmental Conservation's water quality department and lab in Ray Brook. It was fun and I learned a lot about the waterways and watersheds in the region. I ate my lunch each day in the cafeteria at work where one colleague, who was about my mom's age, often had lunch, as well. She was very nice, and we often chatted about various topics during our meals together. But our friendly relationship changed suddenly in the summer of 1990, when Mohawks from the community of Kanesatake near Oka, Quebec, halted a development project that was intended to greatly expand a local golf course.

The Mohawks objected to the development because the area to be cleared was on disputed land and was the resting place of deceased ancestors. The developer and mayor of Oka had no desire to reach an agreement with the Mohawks and began cutting the pine forest there. The people of Kanesatake blocked the road to protect the trees. Canadian police charged in with tear gas only to have the wind shift and blow the gas back into their faces. A gun battle broke out, resulting in the death of a police officer and triggering a prolonged standoff between the Mohawks and the Canadian army. The standoff led to supporting blockades in Kahnawake, and other Indigenous people from elsewhere offered support, as well. Four thousand Canadian troops, as many as were sent to fight in Iraq that year, encircled several dozen Mohawks with barbed wire, armored personnel carriers, and helicopters. Many protesters were beaten and arrested before the conflict ended.

One day the Oka crisis came up in conversation during lunch with my colleague at work, and she voiced an opinion that made my heart drop.

"The army should just drive up to the blockade and mow down those savage Mohawks."

I sat in shocked silence, unable to believe what I had just heard. She didn't know that I was one of those "savage Mohawks."

I rose, walked out, and never returned to that lunchroom again.

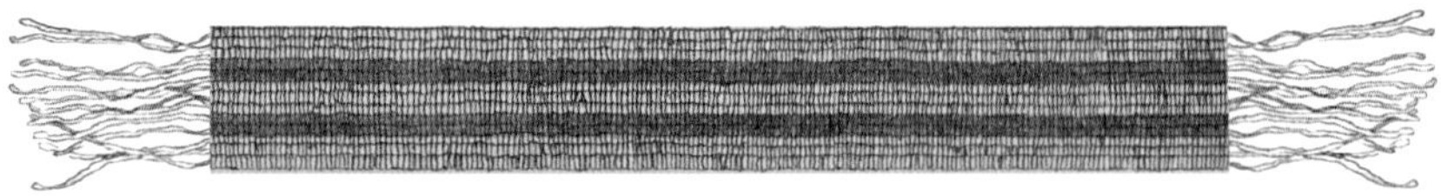

My Haudenosaunee ancestors were tolerant, generous, and hospitable people. Our way of thinking was attractive to the fledgling American colonies, because our society valued freedom of expression and participation. Our political structure is the oldest participatory democracy in the world. Everyone has a voice. Everyone. The founding fathers of the United States were familiar with the Haudenosaunee, and there is evidence that Haudenosaunee political structure was a model, in part, for the United States Constitution. There is scholarly debate as to how much influence there actually was, but when two cultures meet and interact there is bound to be some transference and borrowing between them.

We Haudenosaunee have maintained our cultural and political identity despite decades of war, land loss, and cultural assimilation. Today, the struggles our ancestors endured and the more recent work of visionaries such as my grandfather, Tehanetorens, have encouraged my own generation to openly practice our ceremonies, speak our language, and maintain our rights as sovereign peoples. We now have the legal rights and ability to teach our language in our schools, which was formerly forbidden by law. We can address the United Nations as a sovereign people, and our longhouse is packed during ceremonies. We have become lawyers, teachers, and doctors, always keeping who we are as citizens of the Mohawk Nation and Haudenosaunee Confederacy at the forefront of our identity. This position of empowerment that we now hold also allows us to turn our attention to the land.

Akwesasne straddles the border between Canada and the United States along the Saint Lawrence River. When the seaway and hydropower generation came to the area during the 1950s, dredging of the river and development of shipping locks disrupted the natural movements of fish. At the same time, major polluting industries set up shop along the river to the west of Akwesasne to take advantage of the inexpensive transportation and electric power.

My mother grew up on Kawehnokeh, also known as Cornwall Island, and she remembers that the water used to be so clear that she could drink from the river itself. This is no longer the case. Pollution from the General Motors, Alcoa, and Reynolds Aluminum plants drifted in on the air and in the river water, and pulp mills on the north side of Akwesasne spewed foul-smelling fumes from tall smokestacks. I remember noticing a metallic taste in the air as I walked along the banks of the river while living at my grandparents' house on Kawehnokeh. It was inescapable.

During the late 1970s and 1980s, Mohawk people noticed that their cows' teeth were falling out and many of the fish they caught had skin lesions. Something was wrong. A snapping turtle who was found dead at Raquette Point in Akwesasne was sent to Albany for testing. Ward Stone, a wildlife pathologist for the New York State Department of Environmental Conservation, found the turtle to be loaded with polychlorinated biphenyls (PCBs) and other contaminants at concentrations hundreds of times higher than what is safe for human consumption. Health alerts were sent to those who lived near the river, warning them not eat the fish or any other foods gathered near the river. This led to a drastic change in the lives of many residents of Akwesasne, who had lived off the land and waters there for generations.

When I was eight years old, my brother and I stopped by a favorite swimming hole at Akwesasne while my family visited relatives there. There was a small bay along the river where the current was slow, and kids could easily swim there on hot days. We decided to join the other kids at the swimming hole and dove in, but soon after splashing around in it my brother grabbed me by the arm and told me to get out of the water. He pointed to the other side of the bay where men wearing contamination suits were digging along the shore. We later learned that this spot was a toxic waste dump loaded with PCBs, one of the EPA's top ten cleanup sites in the country. I never swam in the river again.

The snapping turtles who live in the river in and around Akwesasne, and who serve as reminders of our traditional connection to Turtle Island, are now toxic. Harmful concentrations of fluoride and mercury have been found in local cattle, and PCBs have even been found in women's breast milk. Our farms have closed down, and we still receive notices reminding us not to eat more than one fish a month. As people began consuming more of the unwholesome processed foods available in grocery stores instead, the change in diet led to an uptick in diabetes and other ailments.

The Mohawk people knew this landscape and its waterways for countless generations. Since time immemorial, our ancestors depended upon the gifts that the land and water provided us. Now that relationship is gone.

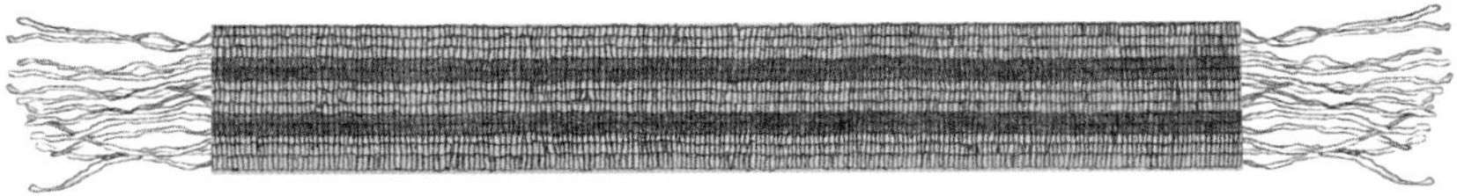

The concept of owning land is foreign to most traditional Indigenous people. I remember hearing one elder compare it to owning air. It is therefore not surprising that most Native people of the past had a hard time understanding the concept when negotiating treaties and agreements with the colonizers. They were more likely to consider such arrangements to be agreements of friendship and cohabitation, and each land exchange brought with it problems of interpretation, understanding, and legality. Today, Native peoples understand that many of the former land transfers were deceptive, and we are now seeking ways to recover our ancestral territories and relearn how to live with the land, not dominate and exploit it.

One approach is to take the land back by occupying it. I remember going with my grandfather to Ganienkeh, an Indigenous settlement where he would often visit and teach the youth. Its story began in 1974, when a group of Mohawks from Kahnawake and Akwesasne decided to bypass the American courts and directly reclaim some of our original territory. They drove to Moss Lake in the Adirondacks between Inlet and Old Forge where a former Girl Scout camp lay vacant. The caravan moved in, took up residence, and renamed the place Ganienkeh, which translates to "Land of the Flint."

The locals were not sympathetic, and they urged the state to evict the intruders. Law enforcement decided to maintain a presence on the sidelines, thinking that this action would be similar to a "sit-in" and that the Mohawks would go home soon. Instead, they held the site for three years.

After much negotiation with then–Secretary of State Mario Cuomo, the people agreed to relocate to Altona, just outside the Adirondack Park. They are still there to this day. The community is dedicated to maintaining and nurturing our culture and making decisions based on Kaianarekowa, the "Great Law of Peace" that underlies the political and cultural structure of the Haudenosaunee nation.

More recently, a global-scale grassroots movement called "Land Back" has sought the return of land to Indigenous peoples who have lost culture, sovereignty, and territory to colonization by European countries. Various groups favor different approaches to Land Back, which include decolonization, environmental restoration, or reestablishing treaty rights, but always with an emphasis on language, traditions, and the teaching of ecological knowledge. It is seen as a way to reconnect with the land and to relearn our roles as caretakers for future generations. At a minimum, Indigenous people want to have a seat at the table when decisions concerning land use are made and to speak on behalf of the earth, land animals, fish, air, and waters.

Land Back actions in recent years have sometimes led to conflict. The Oka crisis, the Wet'suwet'en pipeline protest, the Mi'kmaq lobster dispute, and the Standing Rock protest are just a few examples. Most of these actions have been to defend land against pipelines, logging, damming of rivers, and other activities that damage the earth. In many cases, the corporations responsible have had the support of local government and used armed force in response.

Among the Haudenosaunee and other Native peoples, diplomacy and dialogue are the preferred ways of avoiding violent conflict. We sit in council with other parties to try to find common ground, make concessions, compromise, and perhaps most importantly, to listen. But when we are not even offered a seat, there are few options left for us. For many years Indigenous people in the United States have tried to use the legal system to make the government uphold treaty rights and return land that was unlawfully taken, but for the most part without success. As my grandfather used to say, "No Indian will ever win in a White man's court." Sadly, that has been proven true all too often.

In 2005 the court case known as *City of Sherrill v. the Oneida Nation* was centered on land that was purchased by the Oneidas and deemed to be their tax-free sovereign territory. The city of Sherrill, New York, sued for lost tax revenue, but lower courts ruled that this was indeed sovereign Native land and therefore not taxable. Unfortunately, the US Supreme

Court overturned the lower court's decision, declaring that the Oneida had waited too long to reacquire the land and would therefore put undue hardship on local towns and non-Indigenous residents of the area.

That decision set a precedent for land claim lawsuits across the entire country, which now require more difficult negotiations with state and local municipalities. As a result, alternative means of getting land back are needed, and our museum in Onchiota offers one example of how to do that.

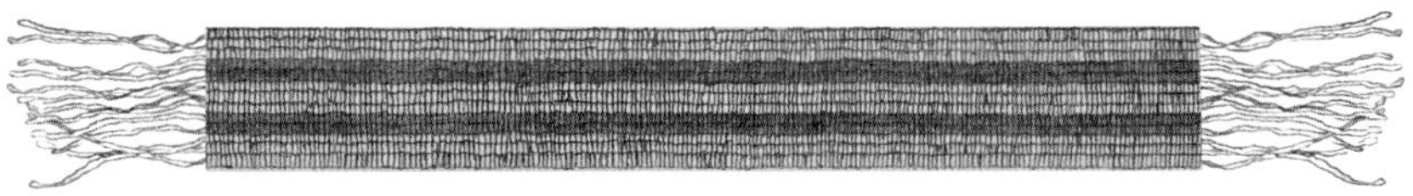

During the early 2000s, I began to share the duties of running our small museum with my dad. I worked at the reservation newspaper, *Indian Time*, and at the Native North American Traveling College, which gave me time off during the summer to keep the museum open. This allowed me to hone my skills as a storyteller and lecturer and also gave my dad a break after spending every summer at the museum since its inception in 1954. Around 2010, I took charge of the center and have spent my summers there ever since.

Interest in Native peoples and our perspectives on relationships with land and the earth have been on the rise in recent years. Climate change due to human activity has caused some folks, especially young folks, to look to Indigenous people for advice and new ways of looking at the relationships we all have with our common home.

During the summer of 2013, Stephanie Ratcliffe, director of the Wild Center in Tupper Lake, stopped by our museum and introduced herself. During this visit, we talked about the possibility of developing an exhibit at the Wild Center that would have a Native perspective on relationships with the land.

From the outset, I envisioned a space in which there would be no sharp edges and that was filled with images of all the elements of Ohenton Karihwatekwen, the traditional thanksgiving address. During one meeting at the beautiful facility, Steph told me to pick out a spot for the installation. A bit taken aback, I selected the oxbow exhibit, which had an indoor pond inhabited by fish, turtles, and ducks and was illuminated by windows that brought natural light in from the sky. A grant was secured, and the hard work of implementation began.

With the support of Wild Center staff and an outside company, the exhibit opened in May 2018, and it still welcomes visitors today. To me, this experience embodied the true meaning of the Two Row Wampum belt. The Wild Center was in their canoe with their natural history and scientific knowledge, and I was in the other canoe with my knowledge as an artist, storyteller, and Indigenous person. This relationship opened my eyes to the wonderful possibilities that can arise when people join together with a common goal, good minds, and a willingness to learn from one another.

When the COVID pandemic took hold of the world in 2020, my family and I had to close the museum for the first time since 1954. The disruption of normal life gave us a chance to think about the future of the collection. The original four-room building lacks running water and bathrooms (we rent a portable toilet every summer) and has no climate control at all, which has led to the degradation of several artifacts that had to be removed for safer storage. My family and I talked about this for years but never had a solid plan other than we knew we needed funding for renovations or an entirely new structure. A retired couple and longtime friends of the family indicated that they would like to help us financially, but we also needed to obtain additional funding through grants.

We decided that it was now or never. My dad contacted a lawyer we had met the previous summer to ask for assistance with becoming a not-for-profit organization with tax-free status that could help us obtain grant support. After many virtual and in-person meetings we filed our paperwork and officially became the Six Nations Iroquois Cultural Center, Inc.

During the summer of 2021, I learned that my cousin, Denise Meyer, intended to sell 300-plus acres of land that she had inherited from her parents, Carroll and Gloria Fadden. Carroll, my paternal uncle, had inherited the bulk of my great-grandfather's acreage and settled on it after retiring from Crucible Steel in Syracuse. They had run a small snack bar there for many summers, a place that all the kids in the area loved for its ice cream and video games. Although my aunt and uncle have now passed away, I always regard that site with fondness. When Denise made the difficult decision to sell the property, I looked for ways in which our new not-for-profit organization could purchase it.

That same summer, a woman named Mary Thill stopped by the center for a visit. She said that she worked for the Adirondack Land Trust and explained that they purchase land for preservation and conservation in the Adirondacks. I replied, "I know where there is some nice land for sale."

After several meetings with the land trust and much negotiation about price, we came to agreeable terms. I would have to raise just over $100,000 and the land trust would pay the other half. The cultural center would own the land and there would be a conservation easement on most of the property. The easement was designed with a strong focus on traditional Indigenous land usage, and its language reflects Haudenosaunee philosophy with respect to the land and environment. As with the Wild Center exhibit, we have approached this partnership as our ancestors did with the Two Row Wampum, as two peoples and two cultures with friendship and mutual aid as a premise.

As if the stars were aligned, more people came out of the woodwork to offer help. In November 2021, I received a call from my cousin Nelson Jock from Akwesasne, who asked if we could open the center and talk with a writer he knew. I agreed and drove to Onchiota to welcome Nelson and his guest. To my surprise, it was Charles Mann, author of the acclaimed book *1491*. He was working on an article for *National Geographic* and had someone with him to take photos for the magazine.

I put my storytelling hat on and gave a presentation about the Kaianerekowa, or Great Law of Peace, the foundational law of the Haudenosaunee Confederacy. After the presentation and a look around the center, we visited for a short while in the chilly late fall air. My brother told Nelson's brother, Howie, about our plans for a new building and mentioned that we needed an architect. Charles then leaned over to my brother and said that his wife, Ray, was an architect who has done work with other Native museums. Soon after, plans for a new building were taking form.

I did not want to lose the comforting feeling of the old building and certainly did not want it to be destroyed or abandoned. At this point in its existence, the original building itself is considered to be an artifact. Ray drafted plans for a new green-tech structure that is 60 feet wide by 200 feet long with the original building embedded in the heart of it. New exhibits will be installed, but visitors will still be able to explore the old building and see my grandfather's work. The new structure will also feature rotating exhibits and modern art, and it will most certainly have a bathroom with running water!

As I pondered ways of finding the necessary funds, I received an email from a longtime friend who sometimes purchases art from me. This time, he commissioned me to paint a portrait of his beloved dog and, as a starving artist, I readily agreed. Over the winter and into early 2022, I attempted to capture the character of this beloved family member for him. I completed the painting and arranged to meet him in Onchiota where he could pick it up. He loved the painting, and as we visited, the topic of our project came up. He told me to let him know when we had a price nailed down and said that he would see what he could do to help.

The appraisal was ready in early spring, and I sent him a note to let him know that we had the price. After several days, he sent an email asking if I could meet him in Onchiota again. When I greeted him at the door he reached into his pocket and handed me a check. After I picked my jaw up from the floor, I knew that my fundraising for the land purchase was over.

In October 2022, the new site became ours. We can now be assured that for coming generations of Native and non-Native people alike, this Adirondack land will be protected from development and serve as a classroom for relearning how to live sensibly within our environment and common home. We share a belief with the folks at the Adirondack

Land Trust that our responsibility is to protect, conserve, and steward the land we walk on.

As we journey down this new path, I feel so much gratitude and appreciation for the folks I have met in recent years who have been so willing to help and to learn about our people and our mission as artists and educators. Despite the negative experiences I've had in my lifetime, I've found that when we travel the waters of our lives together with mutual respect and friendship, many great things can be accomplished for the betterment of the earth and all who live on her.

When our new cultural center is built and the old museum is safely tucked inside, I will be able to look east to Mohawk Mountain and tell my grandparents that the work they did for the benefit of all people is safe for the generations still to come.

APPENDIX

Misinformation about Indigenous history in the Adirondacks has deep roots, it continues today, and it is harmful. Being incorrect is no crime in and of itself, of course, and we know that even our own book might contain some factual errors despite our best efforts. The key to good scholarship is to correct such mistakes when better information empowers—and therefore requires—us to do so. We hope that this appendix can help you recognize and replace long-standing errors in local narratives about the human presence in the uplands if you encounter them, and thereby support more accurate and inclusive stories of what it means to be an Adirondacker.

Many verbal and written accounts of Adirondack history still propagate the false claim of absence, and a reader can easily be misled by them. *The First Adirondackers* can serve as a practical tool kit for challenging such misconceptions in detail, as we shall now demonstrate. Let's build upon the recent podcast excerpt that was presented in the Introduction as the starting point for a deeper dive into the problem.

A major propagator of the opinions expressed in that podcast is an outdated text that was published more than a century ago. The two-volume *A History of the Adirondacks* was written by an amateur historian named Alfred L. Donaldson, and it is still widely praised by those who have yet to acknowledge the flaws in it. It is rife with errors regarding other topics in addition to this one, but examples of inaccurate statements regarding Indigenous history include the following:

> *Indians never made any part of . . . the Adirondack Park their permanent home.*

It must be remembered that the settlement could only have been a summer encampment, or headquarters during the hunting season, vacated of necessity during the long and severe winter.

Many of the relics found in the Adirondacks indicate the presence of a people antedating the Indians and possessing a skill in the rude arts far ahead of theirs.

To evaluate these statements rigorously, we must first consider their source. Donaldson was a banker, not a qualified historian, and he lived at a time when most of the information we now have access to was unavailable to him. He based his writings on a relatively small number of sources who themselves lacked sufficient data about the long human history of the Adirondacks to make many valid conclusions about it. The information we have summarized in the preceding chapters now paints a very different picture, and it is worth taking a moment to practice using it.

Consider, for example, the many and varied items that reveal an endemic human presence in the uplands. The stone implements from Lower Saint Regis Lake, Tupper Lake, Algonquin Lake, the Fulton Chain, and elsewhere confirm that people were here for thousands of years. So, too, do countless tiny flakes of imported chert that lie scattered on lakeshores and popular camping sites, each of which bears silent witness to the production or reshaping of stone implements through the ages. Pottery use at Silver Lake Mountain, Saint Regis Mountain, Rainbow Lake, and many other sites is more suggestive of domestic family life than a few hunters in a hurry. And the dugouts from Twin Ponds are strong evidence of sustained presence, too.

Donaldson's comment about finding items far beyond the capabilities of Indigenous people to produce implies that those people have no deep history in the Adirondacks. It also seems to suggest the existence of some previous lost civilization. Who might those mythical Adirondack predecessors have been? Ancient astronauts or colonists from Atlantis? Far more reasonable conclusions to draw from those artifacts are that Donaldson's source mistook modern items for ancient ones, that Donaldson misunderstood his source, or that the report was simply fabricated.

In light of the guest podcaster's speculation that people challenge the claim of absence only for emotional reasons, it is worth noting that emotional reactions are in fact appropriate in this case. Indigenous residents of the region with whom I've spoken about such claims have told me that they are not only surprised but also hurt, angry, or frustrated by them. How can it not be painful to have your heritage falsely erased like that?

But when it comes to Donaldson, the bigotry in his writings is so extreme, even for the 1920s when White supremacy was more casually overt in the American mainstream, that it calls into question his scholarly judgment, as well. Consider these statements in reference to the racist serial murderer Nathaniel Foster.

> *He had one pet aversion—Indians. In those days a number of those roved through the woods.*
>
> *There is little doubt that he was not always prone to distinguish between a red man and other dark-skinned animals.*

And in describing a so-called half-breed individual, Donaldson quipped, "He could . . . shoot a deer, hook a trout, or smell an Indian quicker than any white man."

A History of the Adirondacks aimed even more of this kind of false, racist invective at early Black settlers of the region, which we need not repeat here. The point is that such words are bright flags that warn of misguided worldviews whose continued presence in our local literature is a discredit to the Adirondack community. Unfortunately, a century of uncritical copycat writing has helped make Donaldson's accounts a persistent canon of regional history.

The aforementioned podcast is just one of many recent manifestations of the problem. For instance, the 1986 book *Peaks and People of the Adirondacks* lauded Donaldson's tome as "the first and only great history of the region . . . that is standard . . . by virtue of its having been composed so accurately, so affectionately, and so well." And a recent opinion piece in the *Adirondack Daily Enterprise* stated, "The finished product of

Donaldson's decade of research and writing is a remarkable and pleasantly readable two-volume masterpiece."

But enough of Donaldson now. We have more ground to cover. Having established on these pages that the old claim of absence is indeed false, let's examine some of the equally flawed arguments that seek to explain why people could not have lived here.

Three of the most commonly proposed rationales are the following: It was too cold, the uplands were unsuitable for agriculture, and the Indigenous people were only passing through. We will now address each of them in turn along with examples that appear in widely used literature. We identify such writings by name here only to show how well-known and influential such sources are. Most of the contents of those publications are sound and enjoy well-deserved reputations for quality, and it is only their comments regarding Indigenous history that we will be challenging.

Question #1: Did winter cold prevent Indigenous peoples from living in the uplands?

Example: *The Adirondack Atlas: A Geographic Portrait of the Adirondack Park* (2004), stated that "the inhospitable environment was avoided by Native Americans and Europeans alike."

Answer: No.

It is not true that the Adirondacks were too inhospitable for non-Europeans because of winter cold. The claim also makes little sense. Why presume that Indigenous people of the past never figured out how to dress properly, how to build a shelter or find food in winter, and how to deal with snow on the ground or even turn it to one's advantage? Who, after all, invented the laced-frame snowshoes that were so widely used by Euro-American folks on the American frontier?

The written record of Father Jogues's winter captivity among Haudenosaunee men, women, and children in the uplands during the 1600s clearly demonstrates otherwise. So, too, does the eyewitness account of Carlos Whitney, who described seeing an Indigenous woman ice-fishing on Upper Saranac Lake during the 1800s.

Direct physical observations of winters in the uplands today likewise challenge the exclusion-due-to-cold arguments. Most of the Adirondack Park, about two-thirds of it in fact, lies between 1,000 and 2,000 feet elevation, and that is where most of the signs of human presence have thus far been found. However, the winters there are not much harsher than in the surrounding lowlands. When freezing temperatures prevail at 1,500 feet around Tupper Lake or 1,700 feet around Indian Lake, it is likely to be only a few degrees warmer down in Canton, which lies at 380 feet elevation in the Saint Lawrence Valley. Winter temperatures at those upland sites averaged 17°F during the past century while the mean in Canton was 18°F, and sometimes the uplands can even be warmer than the lowlands when weather inversions occur.

Now consider those minor differences in practical terms. One can just as easily freeze to death at 18 degrees in the Saint Lawrence Valley as at 17 degrees in Tupper Lake, but nobody is claiming that the cold prevented Indigenous people from living in the Saint Lawrence Valley. It just doesn't add up.

Human physiology also argues against claims of absence due to climate. The effects of cold on the human body are not merely a function of temperature but also of wind. Windchill steals body heat through convective cooling and by evaporative cooling of exposed skin, such that a moderate breeze at 35°F drops the effective temperature below freezing. Residents of frigid places have long known how to dress in ways that maximize insulation and minimize windchill. They have also known how to use the landscape when dealing with winter. Mountains and thick forests help block the force of wind, so there is potentially more shelter from bone-chilling blasts on a rugged Adirondack landscape than in the open lowlands.

Finally, cold and snowy mountain environments can provide logistical advantages that warmer settings lack. When lakes, wetlands, and rivers freeze, it becomes easier to cross the landscape on foot. Frozen waterways can serve as convenient highways without troublesome detours or a need for watercraft. Travel in snow leaves tracks and therefore reduces the risk of becoming lost while also creating more readily visible records of animal presence in the area. Deep drifts can help slow prey down while a hunter pursues their quarry on snowshoes. Cold helps preserve perishable food items, and snow itself can serve as a building material when emergency shelter is needed or a drafty dwelling needs extra insulation packed around it.

That said, we don't really need climatic analyses to demolish the claim of absence on the basis of cold. It is well known that First Nations peoples of this continent have lived—and continue to live—all the way north to the frigid rim of the Arctic Ocean. Undeniable signs of Indigenous presence across the Northeast also date back to the end of the last ice age when winters were even colder than now. If those other regions were occupied for millennia under even harsher conditions, then why suppose that people avoided the Adirondacks because they couldn't handle the climate?

In short, pronouncements that Adirondack winters were too brutal for Indigenous people to endure are firmly and demonstrably wrong. Strike one for the false claim of absence.

Question #2: Did poor agricultural conditions prevent people from living in the uplands?

Example: *Contested Terrain: A New History of Nature and People in the Adirondacks* (1997), reported that "Indians routinely passed through the Adirondacks . . . (due to) characteristics of climate and terrain that discouraged Native Americans . . ." by preventing them from growing crops in the uplands.

Answer: No.

First, let's remember that suitability for farming is irrelevant to most of Adirondack history. Of the 12,000 or more years during which people occupied the uplands, only the last millennium saw fully domesticated plants such as the Three Sisters become major staples of diets in the North Country. That's less than one-tenth of human history in the region.

Yes, many Indigenous peoples of the North Country in recent centuries developed large farming communities in the lowlands where soils tend to be more fertile and climates slightly milder. Among the most notable examples were the fortified settlements of the Haudenosaunee, but many of the Wabanaki peoples who also lived in the region were less dependent upon crops. For many of them, horticulture was more of a dietary supplement than a core requirement of life, so soil quality was not such a crucial factor in their choice of home ranges. Even many Haudenosaunee farmers continued traditional mobile hunting-gathering practices between growing seasons.

In any case, the point is moot. It actually was—and still is—quite possible to harvest crops successfully in the uplands. For one thing, there were many more food plants to consider than the Three Sisters alone. Long before those crops arrived, Indigenous peoples of the Northeast consumed the seeds of wild to semi-wild plants such as goosefoot, sumpweed, little barley, wild rice, and pickerelweed. There were many kinds of fruit including strawberries, raspberries, blueberries, shadberries, elderberries, and chokecherries, all of which could be dried and stored for later use. Jerusalem artichokes, trout lilies, ground nuts, and cattails had edible tubers. Oaks, beeches, and chestnuts provided nuts, and various herbs offered fresh greens. The geographic ranges of such plants extend well into parts of Canada where growing conditions are comparable to or harsher than those in the Adirondacks, so horticulture and plant use writ large were clearly possible in the uplands for thousands of years.

In addition, the people who later established the Three Sisters in the region developed fast-growing, cold-tolerant varieties to yield good harvests in the North Country. Despite generalizations about the poor quality of upland soils, local farmers have long known that sizable pockets of fertile, fine-grained soil were left behind by deglacial lakes. Examples include today's large, productive farm fields in North Elba,

Vermontville, and Gabriels, not to mention potentially fertile floodplains of the region's many rivers and the maize fields at Indian Carry and Piseco Lake during the 1800s.

Local census records show that non-Indigenous people have also grown maize in upland farm and garden plots throughout the last two centuries. One-third of the farms listed in the 1855 agricultural census for St. Armand, for example, were growing significant amounts of maize along with potatoes, wheat, and other crops at roughly 1,500 feet elevation.

If any doubt remains, take a careful look at the uplands today. Here and there you will see maize tassels waving above the squash blossoms in backyard gardens. Better yet, check out the thousands of them that wave every summer in the "Great Adirondack Corn Maze," a popular tourist destination just off the main road in Gabriels.

Strike two for the absence arguments.

Question #3: Did Indigenous people "only pass through" the Adirondacks?

Example: *The Adirondacks: A History of America's First Wilderness* (1997) said of the Adirondack uplands, "The bulk of it was Mohawk territory. This doesn't mean these original (*sic*) proprietors ever lived in the Adirondacks on a full-time basis."

Answer: No.

In addition to mislabeling the Haudenosaunee as "original proprietors" and overlooking the long presence of other cultural groups in the Adirondacks, this quote echoes the podcast's claim that Indigenous people merely crossed quickly through the uplands in order to reach their real homes elsewhere. Let's take a closer look at such wording in general and the assumptions behind it.

Take the word "only" that appears in this section's title as well as the aforementioned podcast. It is a potentially value-laden term that can imply that Indigenous people who moved around on large landscapes didn't truly live on them. The term reflects a narrow cultural viewpoint that devalues traditional mobile lifestyles in favor of more recent sedentary ways that are typical of modern Europe and America. Its roots arise from an older, more sinister train of thought that concluded, "They weren't really using it anyway, so we took it."

Much the same can be said of "passing through." To confuse long-term, long-range mobility in the quest for food with a quick passage through the uplands is to mistake a home for a highway. Hunter-gatherers, whose lifestyles prevailed throughout most of human existence, were not simply moving "through" their landscapes. They were moving upon them, within them, and in deep personal connection to them.

Kerry Hardy, whose book *Notes on a Lost Flute* celebrates Maine's Wabanaki heritage, clearly sums up the situation when he writes that "we might well question the long-held assumption that a fixed-residence agrarian lifestyle is somehow higher on the evolutionary scale of civilization than the mobile food economy practiced by the Wabanaki. European notions of land ownership and management confined my (own) ancestors to small farms, whereas the Wabanaki collectively managed and harvested the whole landscape—in the long run, a much more sustainable strategy."

More simply put, settler lifestyles based on European values kept people in small, enclosed homes, whereas traditional hunter-gatherers lived in huge, open-air homes.

The burden of proof for the claim of absence actually lies with those who believe that early peoples would avoid a vast, resource-rich region like the Adirondacks even though mobile Indigenous groups inhabited similar regions across North America. Think of the Nuche (Ute), Siksika (Blackfoot), and Apsáalooka (Crow) nations of the Rocky Mountains, in addition to the aforementioned Wabanaki who lived in many mountainous areas of New England and the Canadian Maritimes. There is no plausible reason for the Adirondacks to have repelled the kinds of people who knew how to live in such habitats, are known to have occupied the same kinds of places elsewhere, and required large areas to sustain themselves.

What about historical documents that seem to support the absence claim? Such writings represent only the last five centuries of a much longer story, and they must be interpreted carefully. As Melissa Otis explained in *Rural Indigenousness*, there were many possible reasons why early Euro-Americans believed that the uplands had always been uninhabited. These included war, plagues, self-interest in finding unoccupied land to exploit, romanticized concepts of untouched nature, sheer fabrications, and a reluctance to venture uphill to see the place for themselves, not to mention the lack of effective census-taking infrastructure. And as for the longer spans of human history before European contact, let us remember that none of the descriptions penned during the last five centuries represent direct observations of the preceding cultures on this continent, much less in the Adirondacks.

Variable perceptions of local geography may also contribute to the problem. For some who now see the region mainly as a modern hiker's playground, "The Adirondacks" refers only to the more challenging terrain of the High Peaks. In fact, most of the park's acreage lies at intermediate elevations where, as we have already noted, the climates are not tremendously different from those of the adjacent lowlands and where resource-rich lake and forest habitats abound.

But these are largely theoretical arguments. Statements that people were "only passing through" must ultimately contend with historical and archaeological evidence to the contrary. To repeat for the sake of practice, the Twin Ponds dugouts were clearly intended for long-term use in one particular location. Fragile clay pots that would be difficult to transport intact have been discovered throughout the uplands. They strongly suggest the presence of women as well as men, which in turn suggests homelife more than hunting buddies in a hurry. And eyewitness accounts show that Indigenous people were present during all seasons of the year. Clearly, the human story here was much more than a rapid transit to elsewhere.

Strike three for the false claim of absence.

Words matter, so as we develop more accurate and inclusive portrayals of Adirondack history, culture, and geography, it is also worth considering some of the language we use in doing so.

Let's examine some of the terms that are commonly used to describe people and technologies of the past. Some are obviously pejorative in nature, including "redskin," "savage," "half-breed," and "primitive." Others have been designed to facilitate academic discussion, but their side effects can be unintentionally harmful, too.

For example, one widely used scholarly summary of the history of New York refers to the people who hunted with fluted projectile points as "early man." Such a label can seem to imply that those folks were not fully human, perhaps some North American version of *Homo erectus*. To be clear about this, note that the timescales under consideration here are much too short for major evolutionary transformations to occur, just a few thousand years out of the quarter-million or more in which our species has existed. Let's remember that all Adirondackers have always been as fully human as anyone alive today.

The projectile points and other items that academics use to identify specific cultures and time frames are typically named after locations where archaeological studies documented them for research purposes. We may therefore speak of Kirk, Barnes, or Vergennes points, but it is important to remember that those labels refer to settlements that were established by non-Indigenous people, not the Indigenous cultures they are taken to represent.

For a more comprehensive guide to responsible writing and speaking about such matters, we recommend a book by the Opaskwayak Cree author, Gregory Younging, titled *Elements of Indigenous Style: A Guide for Writing By and About Indigenous Peoples*.

So now what? Why is it so important to acknowledge the long Indigenous presence in the Adirondacks that we have written a book about it and you, dear Reader, have come this far through it?

On a practical level, it can influence legal recognition of land-use rights and cultural identity. More than two centuries ago, the state of New

York attempted to justify its seizure of large tracts of Haudenosaunee territory by declaring them empty and therefore free for the taking. Therein, once again, lie the roots of "they weren't really using it, so we took it."

Officially designated wilderness areas within the Adirondack State Park are not meant to be inhabited by people nowadays, and laws that help protect them also restrict or prohibit many traditional uses of animals, plants, and minerals there. However, the long human presence in the uplands reminds us that the standard concept of vacant American wilderness is a recent cultural construct that was brought to this continent by foreigners. It can seem to imply that Indigenous peoples didn't live in undeveloped landscapes or, as Donaldson suggested, that they were more like wildlife than human beings. Acknowledging human history in the Adirondacks is therefore relevant to fully informed management of wild lands within the park.

Finally, the emotional and spiritual toll of having one's heritage ignored or corrupted can be profoundly corrosive and devastating, as Dave explains in his closing chapter. To feel the effects of massive intergenerational trauma and then hear the perpetrators or their descendants tell you that it never even happened is something that nobody should have to endure.

May this book help us all achieve a better understanding and appreciation of what it means to be Adirondackers and Americans, and above all help us to become better humans.

GLOSSARY AND PRONUNCIATION GUIDE

Guidance in the pronunciation and spelling of Abenaki terms was provided by Daniel Gauthier Nolett, executive director of the Abenakis of Odanak Band Council. Guidance in the pronunciation and spelling of Haudenosaunee terms was provided by David Kanietakeron Fadden, Elizabeth Karonhisake Fadden, Margaret Lafrance, and Sue Herne of Akwesasne. Thanks to all for their generosity and patience, with my apologies for any errors that might remain.

Abenaki (ah-BEN-ah-kee or ab-en-AK-ee): the westernmost nation of the Wabanaki confederation within what are now also known as northern New England and eastern Canada. It can translate to "People of the Dawn."

Akwesasne (ah-kweh-SAH-sneh): a Kanienkeha place-name for the Indigenous community on the Saint Lawrence River. It can translate to "Where the Partridge Drums."

Algonquian (al-GOHN-kwee-an): refers to a large family of languages and cultures that have been prominent from the North Atlantic coast westward to the Great Plains, including Wabanaki, Mahicans, Cree, Anishinaabe, and others.

Anentaks (ah-NOON-duks): a Kanienkeha term that can translate to "porcupine" or "bark-eater."

Atirutaks (ah-dee-LOO-duks): a Kanienkeha name applied to certain Algonquian residents of the Saint Lawrence Valley. A possible source of the place-name "Adirondacks."

Ganienkeh (gun-YUH geh): a traditionalist Kanienkehaka community near Altona, New York. The name can translate to "Land of the Flint." It can also be spelled as Kanien:ke.

Glooskap (gloo-SKAHP), also spelled "Gluskabe" or "Klosgab": a magical hero from traditional Wabanaki culture.

Haudenosaunee (hoh-dee-noo-SOH-nee or ho-dee-noo-SHO-nee): the confederacy of Indigenous nations known as Mohawk, Oneida, Seneca, Cayuga, Onondaga, and, most recently, Tuscarora. The name can translate to "People of the Longhouse."

Iroquoian (eer-oh-KWOI-an): refers to a large family of languages and cultures that have been prominent from the Saint Lawrence Valley and eastern Great Lakes to North Carolina, including Haudenosaunee, Wendat, Huron, Cherokee, and others.

Kahionhes (gah-HYOO-hes): Kanienkeha name of John Fadden. It can translate to "Long River."

Kahnawake (gah-nah-WAH-geh), also spelled "Caughnawaga": a Kanienkehaka community on what is now also known as the Canadian side of the Saint Lawrence River.

Kaianarekowa (gah-ya-neh-leh-GO-wah): a Kanienkeha term in reference to what is also known as the "Great Law of Peace."

Kanesatake (gah-neh-sah-DAH-geh): a Kanienkehaka settlement at the confluence of what are now known as the Ottawa and Saint Lawrence Rivers in Quebec, Canada.

Kaniatarowanenneh (gah-nya-dah-lo-WAH-neh-neh): a Kanienkeha name for what is now also called the Saint Lawrence River. It can translate to "Big Waterway."

Kanienkeha (gah-nyeh-geh-HAH): traditional name of the language used by the people of the easternmost Haudenosaunee nation, also known as "Mohawk."

Kanienkehaka (gah-nyeh-geh-HAH-gah): traditional name for the people of the easternmost Haudenosaunee nation, also known as "Mohawk."

Kanietakeron (gah-nyeh-dah-GEH-loh): Kanienkeha name of David Fadden. It can translate to "Patches of Snow."

Karonhisake (gah-loon-hee-SAH-geh): Kanienkeha name of Elizabeth Fadden. It can translate to "Searching Sky."

Kawehnokeh (gah-weh-NO-geh): a Kanienkeha name for what is now also called Cornwall Island on the Saint Lawrence River.

Kchitegw (tsay-eh-TOOK), also spelled "Ktsitegwiiak": Abenaki name for what is now also called the Saint Lawrence River. It can translate to "Big River" or "Great River."

Kohserake (go-seh-LAH-geh): a Kanienkeha name for the Adirondacks. It can translate to "The Place of Winter."

Kwenogamak (kweh-NO-gah-MAHK): an Abenaki name for what is now also called Long Lake. It can translate to "At The Long Lake."

Mdawilha (mm-dah-wihl-HAH): an Abenaki term for the waterfowl known in English as loon, companion of the folk hero Glooscap.

Ohenton Karihwatehkwen (oh-HON-duh gah-lee-wah-DEH-kwon): "Words Before All Else," the traditional Haudenosaunee thanksgiving address.

Paskangamak (pahs-kahn-gah-MAHK): an Abenaki name for what is now also called Tupper Lake. It can translate to "Branch Lake" or "Side Lake."

Pitawbagw (pih-touw-BA-koo), also spelled "Bitawbakw": an Abenaki name for what is now also called Lake Champlain. It can translate to "Double Lake" in reference to its proximity to what is now called Lake George, or "Lake Between" due to its position between the Green Mountains and the Adirondacks.

Pokuizasne Nepes or Nebes (poh-kway-zahs-NEH neh-bes): an Abenaki name for what is now also called Lower Saint Regis Lake. It can translate to "Akwesasne Lake."

Ratirontaks (lah-dee-LOON-duks): a Kanienkeha term for former residents of what is now also known as the central Saint Lawrence Valley. It can translate to "bark-eater" or "porcupine."

Shé:kon (SEH-go): a Kanienkeha expression of greeting. It can translate to "Hello."

Six Nations Iroquois Confederacy: an alliance of Indigenous nations who are now commonly known as Mohawk, Oneida, Onondaga, Cayuga, Seneca, and Tuscarora.

Skawennati (skah-WAH-nah-dee): Kanienkeha name of Christine Fadden. It can translate to "The Other Side of the Word."

Tahawus (normally pronounced "ta-HAWZ"): an informal name that was applied to Mount Marcy by author and poet Charles Fenno Hoffman during the nineteenth century. According to author Russel Carson, it is derived from a Seneca term that can translate to "he splits the sky" or "cloud-splitter."

Tehanetorens (deh-han-eh-DOH-lons): Kanienkeha name of Ray Fadden. It can translate to "He Splits The Pitch" or "He Walks Through The Pines."

Tsikariotanakere (jee-GAHL-YOO-dah-NAH-geh-leh): a traditional Kanienkeha name for the Adirondack uplands. It can translate to "The Place of Animals."

Wabanaki (wahw-ba-NAH-kee or wahw-ba-NAH-keh): a confederation of Indigenous peoples within what are now also known as northern New England and the Canadian Maritimes, including the Abenaki, Micmac, Maliseet, Penobscot, and Passamaquaddy nations. It can translate to "People of the Dawnland," and can also translate more simply to "Dawnland" in reference to the territory itself.

Wampumpeag (WAHM-pum-peeg): a Narragansett or Massachusett term that can translate to "String of White Beads Used As Money."

Wawonbadenek (wah-won-BAD-eh-nehk): an Abenaki name for what is now also known as Mount Marcy and possibly some of the surrounding peaks, as well (Carson, 1927). It can translate to "White Mountains."

Wet'suwet'en (wet-SOH-wih-tihn): an Indigenous nation in what is now also known as British Columbia, associated with a pipeline protest. It can translate to "People of the Wa Dzun Kwuh River (Bulkley River)."

CHAPTER NOTES

ICE, SUN, AND STONE

The basic geological history of the Adirondack dome was derived from Isachsen (1981) and van Diver (1985). The history of deglaciation in the Adirondack region is based largely on Franzi et al. (2016), Feranec and Kozlowski (2017), and Parker et al. (2022), in addition to sediment core research by Overpeck (1985), Whitehead and Jackson (1990), Hansford (2014), LeBouef (2014), Rechlin et al. (2019), and Stager et al. (2016, 2021). The estimate of 3–4 mm/year for current rates of uplift in the Adirondack dome comes from Isachsen (1981).

Slow cyclic movements in the tilt and wobble of the Earth relative to the sun have set the pace of climatic history in the Northeast since the end of the last ice age, when northern summers in particular became warmer than before (Mayewski et al. 2004; Marcott et al. 2013; Zhang et al. 2021). Paleoclimate reconstructions most relevant to the northeastern United States reveal a basic pattern of warming from the deglacial period to a maximum roughly 8–7 kiloyears before present ("kyr BP," or 8,000–7,000 years ago), followed by gradual cooling and wetter climates overall (Noren et al. 2002; Ellis et al. 2004; Marcott et al. 2013; Shuman et al. 2019; Zhang et al. 2021). The oft-cited Younger Dryas climate disruption that chilled Europe and Greenland circa 13–11.7 kyr BP seems to have had only mild effects on the interior of the Northeast (Ellis et al. 2004; Tremblay et al. 2024) and is therefore not considered in the text.

Adirondack Vegetational History

Alpine vegetation communities of the High Peaks are thought by some to be remnants of lowland tundra from deglacial times (Carlson et al. 2011), which is consistent with pollen-based evidence that boreal parkland mosaics prevailed in the uplands soon after deglaciation. The sediment core records from which the vegetational history of the Adirondacks was obtained came primarily from Brandreth Bog (Overpeck 1985); Upper Wallface Pond, Lake Arnold, and Heart Lake (Whitehead and Jackson 1990); Bloomingdale Bog (LeBouef 2014); and Wolf Lake (Hansford 2014).

I have estimated the approximate timing of Adirondack deglaciation on the basis of the transition from silt to organic mud in sediment cores. Strictly speaking, that transition doesn't necessarily represent glacial retreat per se, because silt could still be deposited in a lake for some time by a meltwater river or by erosion from the catchment, and a newly exposed lake could also have remained frozen and inhospitable to the planktonic organisms who would later produce much of the organic matter on the bottom. In other words, the date of sediment transition at the base of a core is likely to be somewhat younger than the actual retreat of glacial ice from the lake, and the estimated 13 kyr BP age of overall deglaciation used throughout the book is just that, a general estimate.

More specifically, the calibrated basal age ranges for the organic lake sediments in Jackson's cores, as obtained from the CALIB program (http://calib.org/calib/calib.html), are as follows: Lake Arnold (11.3–10.7 kyr BP), Upper Wallface (13.6–12.0 kyr BP), and Heart Lake (12.8–11.4 kyr BP). For Brandreth Bog the range was 12.6–11.8 kyr BP (Overpeck 1985) and for Wolf Lake it was circa 13.6–13.2 kyr BP (Hansford 2014). Basal ages for Moody Pond and Lower Saint Regis Lakes are shown in Table 1 along with other unpublished sediment and peat core dates from selected bogs and lakes. For more detail on the calibration of radiocarbon dates, see Reimer (2021).

Table 1. Radiocarbon dates obtained by Stager and colleagues from Adirondack study sites mentioned in text.

The raw radiocarbon ages have been converted to calibrated years "BP" (before present) or "AD" with the CALIB program (Stuiver et al. 2022). Two-sigma probabilities are given in parentheses for the most likely calibrated date ranges. Sites with informal names are described as follows: (1) Sporadically dry "Meadow Pond" is located on the Paul Smith's College Visitor Interpretive Center property beside a steep esker ridge that separates it from Long Pond; (2) "Jones Pond Road Bog" and the smaller (3) "Scarlet Bog" are the first sizable bogs encountered on the left and right sides, respectively, of Jones Pond Road when leaving Brighton for Rainbow Lake.

Location	Material	^{14}C age	Cal. Year Age Ranges	Lab Number
Lower Saint Regis Lake	Organic mud (core base)	12,200 ± 90	14,456–13,805 BP (*0.94*)	OS-131489
Lost Pond	Organic mud (core base)	11,160 ± 50	13,168–12,962 BP (*0.97*)	Beta-165774
Moody Pond	Organic mud (core base)	10,000 ± 60	11,655–11,266 BP (*0.89*)	OS-147336
Meadow Pond (PSC VIC)	Peat (core base)	7,830 ± 50	8777–8452 BP *(0.98)*	Beta-235742
Scarlet Bog (Jones P. Rd.)	Peat (core base)	7,683 ± 50	8550–8386 BP (*0.97*)	AA-62634
Jones Pond Road Bog	Peat (core base)	4,920 ± 100	5907–5466 BP (*1.00*)	Beta-103580
Long Lake	Charcoal (hearth)	1,060 ± 20	976–925 BP (*0.89*)	OS-120809
Twin Ponds	Wood (small dugout)	200 ± 30	1644–1694 AD (*0.37*)	Beta-390190
Twin Ponds	wood (large dugout)	260 ± 30	1518–1590 AD (*0.38*) 1620–1672 AD (*0.54*)	Beta-390189
Lake Placid	wood (dugout)	140 ± 30	1671–1768 AD (*0.37*) 1829–1900 AD (*0.32*)	Beta-386173

The Wabanaki name for pitcher plants ("frog shoes") came from Hardy (2009), and the Haudenosaunee name for them ("turtle socks") was from a personal communication with Neil Patterson Jr. of the Center for Native Peoples and the Environment in Syracuse, New York.

The Champlain Sea

The age range of the Champlain Sea has varied in the scientific literature, but most recent analyses of sediment cores and landforms have narrowed the time frame to roughly 13–9 kyr BP (Rayburn et al. 2005, 2007; Cronin et al. 2008; Belrose 2015). Sporadically brackish conditions might have persisted a bit longer than that in what is now Lake Champlain (Robinson 2012; Belrose 2015). The elevation of the Champlain Sea surface was as much as 350 feet above sea level (Robinson 2012, 2017). Evidence that it extended into the Lake Ontario basin was described in Mather (1917) and Franzi et al. (2016).

Salt-tolerant, cold-climate species who inhabited the Champlain Sea included mollusks such as periwinkles, whelks, scallops, and softshell clams; fish such as capelin, cod, and sculpin; waterfowl including eider ducks and murres; and seaweeds such as kelp (Loring 1980; Harington et al. 2014). Whale species included beluga, bowhead, humpback, and finback as well as harbor porpoises, and seal species included ringed, harp, harbor, and bearded seals in addition to walruses (Loring 1980; Harington and Ochietti 1988; Harington et al. 2006; Bouchard et al. 2011). David Franzi, who discovered the bones of a ringed seal in Plattsburgh with SUNY Plattsburgh students (Feranec et al. 2014), has also identified several locations near the northwestern corner of Lake Champlain where shells of blue mussels and *Macoma* and *Hiatella* clams are abundant.

The salinity of the Champlain Sea varied over time as meltwater floods from sources such as glacial Lake Vermont and Lake Iroquois freshened it on occasion (Rayburn et al. 2005, 2007). The fluctuations influenced the animal community of the sea somewhat, but some species tolerated them more than others, as they do in modern times. For instance, beluga whales are sometimes seen in the Saint Lawrence estuary today, possibly representing an ancient relict population (Harington et al. 2006). As a result, some marine species remained in the Champlain Sea even during flooding episodes when freshwater species such as lake cisco and longnose suckers joined them temporarily (Harington et al. 2014).

Vegetation in the lowlands surrounding the Champlain Sea was at first dominated by spruce forests, which are still common in the far north and high elevations today. Other kinds of forest with more hemlock and birch were also interspersed with open tundra-like areas dominated by mosses, herbs, and shrubs, and more white pines moved in as climates warmed and tundra declined (Cronin et al. 2008; Robinson 2012).

Fluted projectile points and other signs of Paleoindian presence have been found on both sides of Lake Champlain, typically on former beaches and terraces adjacent to the former Champlain Sea (Loring 1980). Local peoples who hunted, gathered, and fished there had a diverse community of prey to exploit during the 3,000- to 4,000-year existence of the sea (Robinson 2012).

Land Animals

A foot bone from a brown bear that was found in a sandy deposit in Lévis, Quebec, has been dated to circa 11.5 kyr BP (Harington et al. 2014). The description of the musk ox vertebra from the Elizabethtown area was based on Rayburn et al. (2007).

Ancient remains of caribou have been found in Dutchess Quarry Cave (Orange County, New York) along with bones of flat-headed peccaries and giant beavers (Funk et al. 1970; Steadman et al. 1997). Caribou were also widespread elsewhere in the Northeast during deglacial times (Fisher and Ostrom 1952; Lemke 2015) and were present in New York State until shortly before the nineteenth century (Cringan 1957). Most experts believe that they were preyed upon by Paleoindian peoples in the region, and that herds of caribou were likely to have been intercepted by hunters in familiar locations rather than being followed randomly (Funk et al. 1970; Loring 1980). Tundra caribou wander more widely than forest caribou, but it is not yet known which kinds of caribou were most common in the boreal parklands of the North Country (Robinson 2012). The presence of caribou in northern Maine during the nineteenth century was documented in Hubbard (1879) and Hardy (2009).

Descriptions of the anatomy, diets, and habitat requirements of mastodons and mammoths were based on Yansa and Adams (2012) and Zazula et al. (2014). Radiocarbon ages of mastodon and mammoth remains in New York were from Feranec and Kozlowski (2012). The isotopic study of human remains from Montana that demonstrated the heavy use of mammoth meat in Paleoindian diets was published by Chatters et al. (2024). Debate over the causes of the extinction of mammoths and mastodons in North America remains inconclusive and sometimes contentious (Haynes 2002; Speth et al. 2013; Boulanger and Lyman 2014; Zazula et al. 2014). We have therefore chosen to keep our discussion of the topic brief.

OLDER THAN THE TREES

Clear signs of Paleoindian presence in the form of fluted projectile points have been found in upland sites among the White Mountains of New Hampshire and Green Mountains of Vermont (Loring 1980; Crock and Robinson 2012; Goodby 2021). Paleoindian sites in Jefferson, Colebrook, and Randolph, New Hampshire, and in western Maine lie within the 1,000- to 1,300-foot elevation range (Spiess et al. 1984; Boisvert 1998, 1999) similar to those in the Adirondacks. Such sites are often associated with major rivers, which were important centers of travel, trade, and subsistence for early peoples of the Northeast (Spiess et al. 1984; Kuehn 1998; Abel and Fuerst 1999; Boisvert 1999; Meltzer 2009).

In a personal communication in 2016, Jon Lothrop reported that a Michaud-Neponset fluted point was found at 2,400 feet elevation on the Taconic Crest Trail in Rensselaer County, New York, which further demonstrates the presence of early peoples at upland sites comparable to the Adirondacks. Abundance of Paleoindian sites in the lowlands of New York State was documented by Ritchie (1965).

Examples of collaboration between professional archaeologists and amateurs were based on Deller and Ellis (2001) and Lothrop et al. (2016).

Projectile Points

Specific details about the locations of archaeological finds were omitted from this book so as to help protect the sites. Most of the information about fluted points from Adirondack sites is based on interviews with Dan Seib, Jon Lothrop, and the staff at PAF as described in the text. A photograph of what appears to be a fluted Clovis-style point that was said to have been found in the vicinity of Wells, in the southern Adirondacks, also appears in a book about lake ecology (Tobiessen 2012). If that claim is correct, then it would confirm the presence of people in the uplands as early as 13,000 years ago. Unfortunately, the book's author has been unable to confirm the origin of the photograph, and the amateur collector in Wells, who would be the most likely source, Eugene "Butch" Zoller, passed away before we could contact him.

The age range of fluted Barnes or Michaud-Neponset projectile points is approximately 12.2–11.6 kyr BP, a cultural time frame that experts call "Middle Paleoindian" (Bradley et al. 2008; Gingerich and Kitchel 2014; Lothrop et al. 2016; Sholts et al. 2017a). These types of point are named after reference sites in Michigan (where they are called "Barnes" points), Maine (Michaud), and Massachusetts (Neponset), but they have been found over a wide area of the Northeast (Carty and Speiss 1992; Ellis and Deller 1997). Some fluted points of the Bull Brook–Athens Hill type that were discovered at the Ausable River farm site adjacent to the Adirondacks might be as old as 12,700 years (Robinson et al. 2017). According to Jonathan Lothrop (personal communication, 2023), a broken Paleoindian point that was found near Carry Falls Reservoir in the uplands was made of Esopus chert from what is now Central New York and appears to be roughly 12,800 to 12,200 years old.

The archaeological evidence that is available from the Adirondacks at the time of this writing leads us to begin our story of human presence in the uplands around 12,000 years ago, although people may well have arrived soon after the ice left around 13,000 years ago.

The reasons for adding a flute to a projectile point despite the difficulty involved and susceptibility to breakage during production remain unclear. Hypothetical explanations range from greater bloodletting to artistic sensibilities, and one analysis suggests that the groove can act as a shock absorber that reduces the risk of tip failure on impact (Thomas et al. 2017). Although fluting is generally associated with the Americas roughly 13.5—11 kyr BP (Bradley et al. 2008; O'Brien et al. 2015; Sholts et al. 2017b), some points with flutes placed at the tip rather than the base have been found in Arabia where they were developed independently about 8 kyr BP (Crassard et al. 2020).

Atlatls

The term "bannerstone" that is often applied to atlatl weights stems from a mistaken idea that the stones were mounted on poles in Roman fashion as banners for symbolic or ritual purposes, although nonutilitarian uses of some of the larger or more ornate stones cannot be excluded (Rataul 2006). Tubular bannerstones were excavated at the Nettling Site, a plowed field near Lake Erie in southwestern Ontario (Ellis et al. 1991). The age of the site was said to be 10–8 kyr, and the items found in it were typical of Early Archaic cultures. The earliest bannerstones known thus far from the Northeast are roughly 8–7 kyr old (Rataul 2006). Nonetheless, many experts consider the general lack of older evidence for atlatls in North America during deglacial times to be due to a shortage of firmly dated sites rather than a true sign of absence (Hutchings 2015).

Materials Other than Stone

Archaeologists tend to focus upon durable stone tools such as fluted projectile points, but early North Americans also used perishable materials such as wood, bone, or plant fibers. Fragments of basketry and cordage that have been found at the Meadowcroft shelter, Pennsylvania, and elsewhere in the Northeast demonstrate that people were using botanical materials throughout the Paleoindian period (Adovasio et al. 2014). The bear claw necklace in the opening vignette was inferred from the documented presence of brown bears near the Champlain Sea (Harington et al. 2014).

Hunting Techniques

A review of Paleoindian lifeways in the vicinity of the Champlain Sea by Robinson (2012) also informed the caribou-hunting vignette. The ecology of caribou and their association with snowfields that discourage biting insects in summer have been documented by Cringan (1957), Farnell et al. (2004), and Hare et al. (2004, 2012). Details of the caribou fence scene were based on investigations of circa 9,000-year-old hunting fences found under Lake Huron (Lemke and O'Shea 2015; Sonnenburg et al. 2015). The long-term nature of fence construction by cooperative groups, the use of blinds, and the transportation of carcasses to the main camp for butchering were discussed in Lemke and O'Shea (2015). That source also noted the use of other strategic hunting spots such as river crossings and natural choke points in the landscape, the importance of spring and fall migration seasons, reliance on cached foods in winter, and social protocols that likely reduced conflict over resources among early Adirondackers. The use of nets for capturing large prey was based on Feranec and Hart (2019).

Caribou hunting was part of the hunting and gathering activities of people from those times (Gingerich and Kitchel 2014; Robinson et al. 2017), but little definitive evidence of other foods used by Paleoindian peoples has yet been found in the Northeast (Speth et al. 2013; Gingerich and Kitchel 2014; Lothrop et al. 2016). That uncertainty extends to the hunting of mammoths and mastodons. For example, mastodon bones and fluted projectile points are found close together at the Hiscock site in what is now western New York, and some of the bones and tusk ivory show signs of processing by humans, but it remains unclear whether the animals were hunted or merely scavenged after they died (Speth et al. 2013). A quote from archaeologist Dena Dincauze (1993) summarizes a general consensus that "Paleoindian forest dwellers were generalist foragers, taking large herbivores opportunistically but not risking their futures by disdaining the great diversity of smaller game and vegetable foods around them."

Dogs were kept by early people and were typically of relatively small size, are thought to have served as pets and hunting companions, and were sometime eaten, as well (Power and Peterson 1984; Fritz 1999; Madrigal 2017).

Some investigators have hypothesized that early peoples might have used boats or rafts on the Champlain Sea or inland waterways for long-distance trade, seal-hunting, collecting bird eggs from offshore islands, or fishing (Jodry 2005; Robinson et al. 2017), but firm evidence for watercraft in the North Country prior to the last millennium is currently lacking.

Lodging

The residences that were typically used by the highly mobile peoples of deglacial times were portable or otherwise ephemeral. The lodges in the knoll vignette were modeled on information provided by archaeologist Robert Goodby in his book *A Deep Presence: 13,000 Years of Native American History*. Investigations of the Tenant Swamp Site in Keene, New Hampshire, by Goodby and colleagues revealed oval sleeping spaces 12–16 feet long by 10–12 feet wide that represented the footprints of Paleoindian lodges covered with animal hides, most likely those of caribou. Goodby also noted the use of tipi-shaped, hide-covered tents by Inuit peoples of northern Quebec.

Ancient Origin Accounts

Historical and statistical analyses of origin stories from around the world find similarities among many of them that suggest they arose from a single source in the distant past. So-called Earth Diver accounts similar to the Haudenosaunee account of Creation are widely known in North America and eastern Asia (Reichard 1921; Leeming 2005).

THE MISTS OF TIME

Scarcity of "Early Archaic" Sites in the Northeast

Ellis et al. (1991) noted that some scholars "have argued for a virtual abandonment of the area in Early Archaic times due to environmental constraints . . . whereas others . . . have argued against such a position." However, the presence of people in the Northeast throughout the Archaic period is now well established, and the low numbers of Early Archaic sites are thought to be more the result of failure to locate or recognize them than of exceptionally low population numbers (Chapman 1980; Spiess et al. 1984; Ellis et al. 1991; Sanger 1996).

Geography of the Lowlands

As the land beneath the Champlain Sea rebounded in the millennia after the heavy ice sheet retreated, former beaches rose by several hundred feet above the level of what are now the Saint Lawrence River and Lake Champlain. Mapping by Lewis et al. (2021) positioned the raised beaches in a west-to-east line at roughly 450 feet (140 m) elevation between Saint Regis Falls and Akwesasne and is consistent with the findings of Robinson et al. (2017). The shorelines shown on the map in this chapter were documented recently by David Franzi and colleagues.

Lifestyles

Experts suggest that a wide selection of plant and animal foods was used by early peoples despite the general scarcity of evidence for it among archaeological sites of the Northeast (Kuehn 1998). Gingerich and Kitchel (2014) reported that the use of plant foods seems to have increased with the onset of the Early Archaic cultural period. They noted that fire-cracked rock (FCR) is more commonly found in Early Archaic sites than in the older Paleoindian sites and proposed that the stone-boiling that produced the FCR was likely associated with the preparation of plant foods rather than meat.

Plant foods that would also have been available to early Adirondackers of the Archaic period included cattail tubers, beechnuts and hickory nuts, berries, and chenopod seeds (Gingerich and Kitchel 2014). Animal remains that have been found at late Paleoindian and Early Archaic sites in what is now New York state include those of birds, fish, turtles, freshwater mussels, moose, deer, beavers, and porcupines (Kuehn 1998).

Seasonal availability of foods helped shape the patterns of land use, and several authors have used such information to propose how people moved through the year (Power and Peterson 1984; Kuehn 1998; Madrigal 2001, 2017). Spring was a time when larger groups likely gathered to exploit spawning salmon, shad, alewives, or suckers as well as migrating or nesting waterfowl (Lemke and O'Shea 2015; Webster 2023). Summer brought a diversity of plant and animal foods in and around lakes, streams, and wetlands. The age and physical condition of animals such as fawns and brooding passenger pigeons have also been used by archaeologists to infer seasonal use (Madrigal 2001).

Fire-making

A global survey of traditional cultures demonstrated the universal importance of fire for cooking, heat, light, and ritual purposes; the reliance of many cultures on fire-making experts; and a widespread preference for transporting or sharing embers rather than creating fire from scratch (McCauley et al. 2020). Descriptions of traditional fire-making techniques were derived from Taylor (2009), Sorenson (2019), and McCauley et al. (2020). Paul Smith's College student Alex Comer provided personal demonstrations of hand-drill fire-making and the preparation of conifer bark as famine food.

PEOPLE OF THE LAKES

Winter Vignette Details

Intact atlatl darts that have been found in the Yukon ice fields are as much as 6,000 years old. Some of them have revealed the use of beaver castoreum, spruce resin, and ocher as adhesives for attaching fletching to the dart shafts and for strengthening the shafts themselves (Helwig et al. 2021). The tales of Glooskap reflect the possibility that residents of the Adirondacks 5,000 years ago were forebears of some of today's Wabanaki peoples; ancestors of the Haudenosaunee would not arrive until several millennia later.

Foods Available

Several authors (Kuehn 1988; Madrigal 2000, 2001; Taché et al. 2008; Gingerich and Kitchel 2014) list plant and animal remains that have been found in Paleoindian and Archaic sites of the Northeast, many of which could also have been available in the Adirondacks 5,000 years ago. Those included cattail tubers, various berries, and the seeds of chenopods and water lilies. Animal remains included those of deer, squirrels, passenger pigeons, turkeys, geese, foxes, hares, beavers, porcupines, muskrats, turtles, freshwater mussels, and fish. Gingerich and Kitchel (2014) also analyzed the nutritional contents of botanical foods and found that cattail tubers were among the most calorie-rich items, but only in autumn when the plants have recharged them with photosynthetic starches during the growing season. Information on the use of nets, weirs, and group collaboration for fishing was largely based on Taché et al. (2008).

Speculations about the nature and possible effects of the hemlock decline on Adirondack ecosystems and cultures were based on Oswald and Foster (2011), Fiedel (2014), and LeBouef 2014). LeBouef's analyses from Bloomingdale Bog placed the steepest decline around 4,900 years ago, but we have rounded that date up to 5,000 years ago for more convenient use in our imaginary vignette in this chapter.

CROPS AND CLAY

The tribal nations and confederacies who resided in the Northeast in recent centuries did not exist as such prior to the last millennium (Englebrecht 1999; George-Kanentiio 2007; Hart and Brumbach 2015; Birch and Hart 2018). Algonquian groups, including ancestors of the Wabanaki, were probably present in the North Country for thousands of years (Haviland and Power 1981; Wiseman 2001; Hardy 2009), but details of the history of Haudenosaunee presence remain unclear. Some sources, including Snow (1998), place their first arrival in what is now Central New York around 1,000 years ago. Radiocarbon dating of Iroquoian village sites suggests that such settlements first appeared during the 1400s AD (Hart and Brumbach 2009; Abel et al. 2019), shortly before contact with Europeans. The so-called Saint Lawrence Iroquoian peoples were dispersed for unknown reasons between circa 1520 and 1603 (Pendergast 1991; Abel et al. 2019).

Where the ancestors of the Haudenosaunee lived in previous millennia remains a mystery, but most archaeologists, linguistic analyses, and oral traditions place them farther south or west in the Appalachians or Ohio Valley, respectively (Parker 1916; Speck 1923; George-Kanentiio 2007; Whyte 2007). If those accounts are correct, then the Adirondack region was mainly inhabited by the ancestors of Algonquian peoples until Iroquoian peoples joined or displaced them shortly before the arrival of Europeans.

Bows versus Atlatls

Bows and arrows began to replace atlatls and darts relatively recently, perhaps beginning in the American Southwest about 1,500 years ago and somewhat later in the Northeast (Blitz 1988; Marschner and Mason 2013; Whittaker et al. 2017). Experimental studies demonstrate that arrows tend to fly faster and straighter than atlatl-driven darts and they can be fired with more stealth from more positions, which has led some experts to suggest that warfare was a factor in the spread of arrow use in the Americas (Marschner and Mason 2013). However, the lighter weight of arrows can give them less penetration force, so they are not necessarily more effective for hunting large prey than darts and spears (Whittaker et al. 2017).

Crops

Women are thought to have traditionally been the primary cultivators and developers of crops among Indigenous cultures of the Northeast (Fritz 1999). Gourds may have been used as fishing floats and containers before becoming a major food source (Fritz 1999; Hart et al. 2004; Smith 2006). Processing of gourd seeds to make them edible was described by Hart (2004). Seeds that reflect the domestication of squash date back 10,000 years or so in Mexico (Smith 1997), and fragments of gourd rinds dated to 7–6 kyr BP have been found in Illinois, Tennessee, and Maine, but it was unclear whether those gourds were wild or cultivated (Fritz 1999). More clearly domesticated forms of *Cucurbita* in the Northeast date to roughly 4 kyr BP (Fritz 1999) and roughly 3,000 years ago in Central New York (Hart et al. 2007; Hart 2008).

Maize history was mainly based on Smith (2006), Pohl et al. (2007), Hart et al. (2007), Hart (2008), Hart and Brumbach (2009), Piperno et al. (2009), and Hart and Lovis (2013). Early varieties of maize, including Northern Flint, were grown in what is now New York State 2,300–2,200 years ago and had become common by 1,500 years ago, well before large agricultural longhouse villages were established (Hart et al. 2007; Hart 2008; Hart and Lovis 2013; Saint-Pierre and Thompson 2015). The example of "fast maize" was taken from Hardy (2009). The integration of seasonally mobile farmer and hunter-gatherer lifestyles among the Abenaki was based on Hart and Lovis (2013) and Hardy (2009).

The history of bean cultivation was based on Kaplan et al. (1973), Smith (2006), Bitocchi et al. (2012), and Castro-Guerrero et al. (2016). Their spread into the region about 2,000 years ago and first major abundance around 700 years ago were documented by Hart (2008) and Hart and Lovis (2013).

Other plants showing signs of domestication in the Northeast roughly 4,000 years ago included sumpweed or marsh elder (*Iva annua*), erect knotweed (*Polygonum erectum*), sunflowers (*Helianthus annuus*), and goosefoot or lamb's quarters (*Chenopodium berlandieri*) (Fritz 1999; Smith 2006; Hart and Brumbach 2009; Smith and Yarnell 2009; Hart and Lovis 2013; Wales et al. 2017). The history of tobacco was based on Groark (2010), Oyuela-Caycedo and Kawa (2015), Rafferty (2016), Carmody et al. (2018), and Tushingham et al. (2018). Fully wild grasses whose seeds were commonly used included wild rice (*Zizania palustris*), little barley (*Hordeum pusillum*), canary grass (*Phalaris caroliniana*), panic grass (*Panicum* spp.), and barnyard grass (*Echinochloa* spp.) (Smith and Yarnell, 2009; Hart and Lovis, 2013). Taché et al. (2008) speculated that maple sugar was not widely made from maple sap and syrup until demand from European settlers encouraged the change.

The description of sediment core research at Crawford Lake was derived from Ekdahl et al. (2004) and McAndrews and Turton (2010).

Charcoal analyses of Adirondack sediment cores from Piseco and Wolf Lakes are reported in Grochocki et al. (2019), Kong (2015), and Stager et al. (2016). The historical accounts of Indigenous farms along the Saint Lawrence and Hudson Rivers, and of maize varieties suitable for short growing seasons, were derived from Parker (1910).

Pottery

The history of ceramics in the Northeast, their early use in the processing of fish, and the importance of fishing in addition to hunting were based on Taché et al. (2008), Hart and Brumbach (2009), Craig et al. (2013), Taché and Craig (2015), and Feranec and Hart (2019). Hart et al. (2007) date the first appearance of pottery in what is now New York State circa 3 kyr BP.

The vignette in which Mother trains Daughter to make pottery was based on the discovery of a child's finger imprints on Iroquoian pottery from the 1400s (Dorland 2018). Use of wooden bowls and cornmeal-thickened stew in the vignette was based on Parker (1910).

The discovery of an intact pot in the uplands by logger Ed King was described by Schaefer (2013). Taché et al. (2008) reported that large ceramic vessels were unlikely to have been transported over great distances and were therefore likely to have been cached for repeated use in a particular location. The authors also noted that large vessels were likely used for feeding large groups of people.

THE GREAT DISRUPTION

Large, palisaded villages are thought by archaeologists to have become common in what is now New York State during or shortly before the 1400s in response to warfare with Indigenous neighbors (Ritchie 1965; Kerber 2007; Kuhn 2007). The regionalized ethnic groups who comprise the Six Nations Confederation today are thought to have taken their current forms during or shortly after the 1500s (Englebrecht 1999; Hart and Brumbach 2015). The eyewitness account of Isaac Jogues regarding families wintering in the Adirondacks came from Shea (1856) and Talbot (1935).

The overview of Indigenous watercraft in North America is largely based on Monk (1999). The descriptions of Adirondack dugout canoes were based on personal research by the authors (Stager et al. 2022). The poem by Maurice Kenny was published in Kenny (1995). Details on the dugouts from Florida are from Monk (1999), Wheeler et al. (2003), and Greber et al. (2012). The vignette in which residents of the knoll are displaced to make room for Paul Smith's Hotel is based on Wardner (2010).

The description of an Indigenous woman of probable Abenaki descent ice-fishing on Upper Saranac Lake was reported in Otis (2018).

The summaries of intertribal conflicts, alliances, various treaties, and large land transfers are based on personal communications and written works of Kanienkehaka cultural historian Darren Bonaparte (2024a, b) and the writings of Kelsay (1984), Kuhn (2007), Sulavik (2007), Noble (2015), Amrhein (2016), Hardy (2009), Otis (2018), Sherry (2020), and Sasso (2023).

The quote regarding the recent arrival of Haudenosaunee in the North Country was taken from Bonaparte (2024). Sources for the history of the names "Adirondacks" and "Tahawus" include Emmons (1838), Prince (1900), Speck (1923, 1940), Carson (1986), and Sulavik (2007). The implication that the bark-eater epithet arose from scorn felt by agricultural Iroquoian people toward Algonquian hunter-gatherers is supported by the seventeenth-century account of Jesuit missionary Joseph-François Lafitau in Sulavik (2007). Other sources of information regarding local place-names include Prince (1999), Beauchamp (1907), Carson (1986), Otis (2018), Daniel Gauthier Nolett, and the Fadden family.

Sources for the Abenaki origin and translations of the names "Adirondack" and "Mohawk" include personal communications with residents of Akwesasne and an interview with Abenaki scholar Paul-René Tamburro at the Abenaki Heritage Festival in Vermont in June 2023. In an interview with Prince (1900), Mitchell Sabattis provided traditional Abenaki names for various Adirondack landforms. Information on the ethnic background of Moses "Sangemo" Saint-Germain was provided by descendant Mateo Champlain.

APPENDIX

Harris (2002) and town census records provide extensive documentation of maize and other crops being grown successfully by farmers throughout the Adirondacks during the 1800s. Otis (2018) reported eyewitness accounts of Indigenous people growing maize in the uplands then, as well. The climatic data from the Adirondack uplands were compiled using weather station records from the United States Historical Climatology Network. Discussion of wilderness as a cultural construct was based on Cronon (1995).

REFERENCES

Abel, T. J., and D. N. Fuerst. 1999. Prehistory of the St. Lawrence River headwaters region. *Archaeology of Eastern North America* 27: 1–53.

Abel, T. J., J. L. Vavrasek, and J. P. Hart. 2019. Radiocarbon dating the Iroquoian occupation of northern New York. *American Antiquity* 84: 748–61.

Adovasio, J. M., O. Soffer, J. S. Illingsworth, and D. C. Hyland. 2014. Perishable fiber artifacts and paleoindians: New implications. *North American Archaeologist* 35: 331–52.

Amrhein, C. 2016. *A History of Native American Land Rights in Upstate New York*. Charleston, SC: The History Press.

Beauchamp, W. M. 1907. Aboriginal Place Names of New York, New York State Education Department Bulletin No. 400 (Albany: New York State Museum). Accessed Dec. 10, 2023, at https://archive.org/details/aboriginalplacen00beau/page/80/mode/2up?view=theater.

Belrose, A. T. 2015. The Champlain Sea/Lake Champlain transition recorded in the northeast arm of Lake Champlain, USA-Canada. Graduate College Dissertations and Theses. 349. https://scholarworks.uvm.edu/graddis/349.

Birch, J., and J. P. Hart. 2018. Social networks and northern Iroquoian Confederacy dynamics. *American Antiquity* 83: 13–33.

Bittochi, E., et al. 2012. Mesoamerican origin of the common bean (*Phaseolus vulgaris* L.) is revealed by sequence data. *Proceedings of the National Academy of Sciences* E788–E796.

Blackhawk, N. 2023. *The Rediscovery of America: Native Peoples and the Unmaking of U.S. History*. New Haven: Yale University Press.

Blitz, J. H. 1988. Adoption of the bow in prehistoric North America. *North American Archaeologist* 9: 123–45.

Boisvert, R. A. 1998. The Israel River complex: A Paleoindian manifestation in Jefferson, New Hampshire. *Archaeology of Eastern North America* 26: 97–106.

Boisvert, R. A. 1999. Paleoindian occupation of the White Mountains, New Hampshire. *Géographie physique et Quaternaire* 53: 159–74.

Bonaparte, D. 2024. The Seven Nations of Canada: The Other Iroquois Confederacy. Unpublished manuscript, accessed Jan. 24, 2024 at http://www.wampumchronicles.com/sevennations.html.

Bonaparte, D. 2024a. The Seven Nations of Canada Treaty of 1796 and the Mohawk Nation Treaty of 1797. Unpublished manuscript, personal communication Jan. 17, 2024.

Bonaparte, D. 2024b. The history of Akwesasne from pre-contact to modern times. Unpublished manuscript, accessed Jan. 10, 2024 at https://www.wampumchronicles.com/history.html.

Bouchard, M. A., C. R. Harington, and J. P. Guilbault. 2011. First evidence of walrus (*Odobenus rosmarus* L.) in Late Pleistocene Champlain Sea sediments, Quebec. *Canadian Journal of Earth Sciences* 30: 1715–19.

Boulanger, M. T., and R. L. Lyman. 2014. Northeastern North American Pleistocene megafauna chronologically overlapped minimally with Paleoindians. *Quaternary Science Reviews* 85: 35–46.

Bradley, J. W., A. E. Spiess, R. A. Boisvert, and J. Boudreau. 2008. What's the point? Modal forms and attributes of Paleoindian bifaces in the New England–Maritimes region. *Archaeology of Eastern North America* 30: 119–72.

Carlson, B. Z., J. S. Munroe, and B. Hegman. 2011. Distribution of alpine tundra in the Adirondack mountains of New York, U.S.A. *Arctic, Antarctic, and Alpine Research* 43: 331–42.

Carmody, S., J. Davis, S. Tadi, J. S. Sharp, R. K. Hunt, and J. Russ. 2018. Evidence of tobacco from a Late Archaic smoking tube recovered from the Flint River site in southeastern North America. *Journal of Archaeological Science: Reports* 21: 904–10.

Carson, R. M. L. 1927. *Peaks and People of the Adirondacks*. New York: Doubleday, Page, & Co.

Carty, F. M., and A. E. Speiss. 1992. The Neponset Paleoindian site in Massachusetts. *Archaeology of Eastern North America* 20: 19–37.

Castro-Guerrero, N. A., M. C. Isidra-Arellano, D. G. Mendoza-Cozatl, and O. Valdés-López. 2016. Common bean: A legume model on the rise for unraveling responses and adaptations to iron, zinc, and phosphate deficiencies. *Frontiers in Plant Science*. doi: 10.3389/fpls.2016.00600.

Chapman, J. 1980. The early and middle Archaic periods: A perspective from eastern Tennessee. Research Report 19: *Proceedings of the Conference on Northeastern Archaeology* 10.

Chatters, J. C., B. A. Potter, S. J. Fiedel, J. E. Morrow, C. N. Jass, and M. J. Wooller. 2024. Mammoth featured heavily in Western Clovis diet. *Science Advances* 10, eadr3814.

Craig O. E., et al. 2013. Earliest evidence for the use of pottery. *Nature* 496: 351–54.

Crassard, R., V. Charpentier, J. McCorriston, J. Vocges, S. Bouzid, and M. D. Patraglia. 2020. Fluted-point technology in Neolithic Arabia: An independent invention far from the Americas. *PLoS ONE*. https://doi.org/10.1371/journal.pone.0236314.

Cringan, A. T. 1957. History, food habits and range requirements of the woodland caribou of continental North America. *Transactions, North American Wildlife Conference* 22: 485–501.

Crock J. G., and F. W. Robinson IV. 2012. Maritime mountaineers: Paleoindian settlement patterns on the west coast of New England, pp. 48–76 in C. Chapdelaine (ed.), *Late Pleistocene Archaeology and Ecology in the Far Northeast*. College Station: Texas A&M University Press.

Cronin, T. M., et al. 2008. Impacts of post-glacial lake drainage events and revised chronology of the Champlain Sea episode 13–9 ka. *Palaeogeography, Palaeoclimatology, Palaeoecology* 262: 46–60.

Cronon, W. 1995. The trouble with wilderness; or, getting back to the wrong nature, pp. 69–90 in W. Cronon (ed.), *Uncommon Ground: Rethinking the Human Place in Nature*. New York: W.W. Norton & Company.

Deller, D. B., and C. J. Ellis. 2001. Evidence for late Paleoindian ritual from the Caradoc Site (AfHj-104), southwestern Ontario, Canada. *American Antiquity* 66: 267–84.

Dincauze, D. F. 1993. Fluted points in the eastern forests, pp. 279–92 in O. Soffer and N. D. Praslov (eds.), *From Kostenki to Clovis: Upper Paleolithic–Paleo-Indian Adaptations*. New York: Plenum Press.

Dorland, S. G. H. 2018. The touch of a child: An analysis of fingernail impressions on Late Woodland pottery to identify childhood material interactions. *Journal of Archaeological Science: Reports* 21: 298–304.

Ekdahl, E. J., et al. 2004. Prehistorical record of cultural eutrophication from Crawford Lake, Canada. *Geology* 32: 745–48.

Ellis, C. J., S. Wortner, and W. A. Fox. 1991. Nettling: An overview of an Early Archaic "Kirk Corner-notched Cluster" site in southwestern Ontario. *Canadian Journal of Archaeology* 15: 1–23

Ellis, C. J., and D. B. Deller. 1997. Variability in the archaeological record of northeastern Paleoindians: A view from southern Ontario. *Archaeology of Eastern North America* 25: 1–30.

Ellis, K. G., H. T. Mullins, and W. P. Patterson. 2004. Deglacial to middle Holocene (16,600 to 6,000 calendar years BP) climate change in the northeastern United States inferred from multi-proxy stable isotope data, Seneca Lake, New York. *Journal of Paleolimnology* 31: 343–61.

Emmons, E. 1838. New York Assembly Document No. 200, February 20. Albany, NY.

Engelbrecht, W. 1999. Iroquoian ethnicity and archaeological taxa, pp. 51–59 in R. F. Williamson and C. M. Watts (eds.), *Taming the Taxonomy: Toward a New Understanding of Great Lakes Archaeology*. Toronto: eastendbooks.

Farnell, R., P. G. Hare, E. Blake, V. Bowyer, C. Schweger, S. Greer, and R. Gotthardt. 2004. Multidisciplinary investigations of alpine ice patches in Southwest Yukon, Canada: Paleoenvironmental and paleobiological investigations. *Arctic* 57: 247–59.

Feranec, R. S., D. A. Franzi, and A. L. Kozlowski. 2014. A new record of ringed seal (*Pusa hispida*) from the late Pleistocene Champlain Sea and comments on its age and paleoenvironment. *Journal of Vertebrate Paleontology* 34: 230–35.

Feranec, R. S., and J. P. Hart. 2019. Fish and maize: Bayesian mixing models of fourteenth- through seventeenth-century AD ancestral Wendat diets, Ontario, Canada. *Nature Scientific Reports* 9: 16658. doi.org/10.1038/s41598-019-53076-7.

Feranec, R. S., and A. L. Kozlowski. 2017. Ice age mammals colonize New York: A STEM lab derived from collections-based research at the New York State Museum. New York State Museum *Education Leaflet* 37.

Feranec, R. S., and A. L. Kozlowski. 2012. New AMS radiocarbon dates from late Pleistocene mastodons and mammoths in New York State, USA. *Radiocarbon* 54: 275–79.

Fiedel, S. J. 2014. Abrupt changes of climate and of point styles along the Atlantic seaboard of eastern North America: How were they connected? *Archaeology of Eastern North America* 42: 77–100.

Fisher, D. W., and J. H. Ostrom. 1952. A contribution to the Pleistocene fauna of New York State. *American Journal of Science* 250: 609–16.

Franzi, D. A., J. C. Ridge, D. L. Pair, D. Desimone, J. A. Rayburn, and D. J. Barclay. 2016. Post-valley heads deglaciation of the Adirondack Mountains and adjacent lowlands. *Adirondack Journal of Environmental Studies* 21: 119–46.

Fritz, G. J. 1999. Gender and early cultivation of gourds in eastern North America. *American Antiquity* 64: 417–29.

Funk, R. E., D. W. Fisher, and E. M. Reilly. 1970. Caribou and Paleo-Indians in New York State: A presumed association. *American Journal of Science* 268: 181–86.

George-Kanentiio, D. 2007. Iroquois roots. Ch. 20 in J. E. Kerber (ed.), *Archaeology of the Iroquois. Selected Readings and Research Sources*. Syracuse, NY: Syracuse University Press.

Gingerich, J. A. M., and N. R. Kitchel. 2014. Early paleoindian subsistence strategies in eastern North America: A continuation of the Clovis tradition? Ch. 16 (pp. 297–318) in A. M. Smallwood and T. A. Jennings (eds.), *Clovis: On the Edge of a New Understanding*. College Station: Texas A&M University Press.

Goodby, R. G. 2021. *A Deep Presence: 13,000 Years of Native American History*. Portsmouth, NH: Peter E. Randall.

Greber, N. B., K. C. Ruhl, and I. Greber. 2012. The Ringler dugout revisited. *Journal of Ohio Archaeology* 2: 16–29.

Groarck, K. P. 2010. The angel in the gourd: Ritual, therapeutic, and protective uses of tobacco (*Nicotiana tabacum*) among the Tzeltal and Tzotzil Maya of Chiapas, Mexico. *Journal of Ethnobiology* 30: 5–30.

Grochocki, K. K., C. S. Lane, and J. C. Stager. 2019. An 1800-year record of environmental change from the southern Adirondack Mountains, New York (USA). *Journal of Paleolimnology* 62: 301–14.

Hansford, R. 2014. A Holocene-scale paleolimnogical analysis of climate-related shifts in pollen assemblages in Wolf Lake (Adirondacks, NY, USA). Honors thesis for Bachelors of Science degree, Queens University, Kingston, Ontario.

Hardy, K. 2009. *Notes on a Lost Flute: A Field Guide to the Wabanaki*. Rockport, ME: Down East Books.

Hare, P. G., S. Greer, R. Gottardt, R. Farnell, V. Bowyer, C. Schweger, and D. Strand. 2004. Ethnographic and archaeological investigations of alpine patches in southwest Yukon, Canada. *Arctic* 57: 260–72.

Hare, P. G., C. D. Thomas, T. N. Topper, and R. N. Gotthardt. 2012. The archaeology of Yukon ice patches: New artifacts, observations, and insights. *Arctic* 65: 118–35.

Harington, C. R., M. Cournoyer, M. Chartier, T. L. Fulton, and B. Shapiro. 2014. Brown bear (*Ursus arctos*) (9880 +/- 35 BP) from late-glacial Champlain Sea deposits at Saint-Nicolas, Quebec, Canada, and the dispersal history of brown bears. *Canadian Journal of Earth Sciences* 51: 527–35.

Harington, C. R., S. Lebel, M. Paiement, and A. de Vernal. 2006. Félix: A Late Pleistocene white whale (*Delphinapterus leucas*) skeleton from Champlain Sea deposits at Saint-Félix-de-Valois, Québec. *Géographie Physique et Quaternaire* 60: 183–98.

Harington, C. R., and S. Ochietti. 1988. Systematic inventory and paleoecology of marine mammals from the Champlain Sea and its approaches. *Géographie Physique et Quaternaire* 42: 3–193.

Harris, G. 2002. The hidden history of agriculture in the Adirondack Park, 1825–1875. *New York History* 83: 165–202.

Hart, J. P. 2008. Evolving the Three Sisters: The Changing Histories of Maize, Bean, and Squash in New York and the Greater Northeast. Ch. 7 (pp. 87–99) in J. P. Hart (ed.), *Current Paleoethnobotany II*, Albany: New York State Museum. doi: 10.13140/2.1.1677.5362.

Hart, J. P. 2004. Can *Cucurbita pepo* gourd seeds be made edible? *Journal of Archaeological Science* 31: 1631–33.

Hart, J. P., and H. J. Brumbach. 2009. On pottery change and northern Iroquoian origins: An assessment from the Finger Lakes region of Central New York. *Journal of Anthropological Archaeology* 28: 367–81.

Hart, J. P., H. J. Brumbach, and R. Lusteck. 2007. Extending the phytolith evidence for early maize (*Zea mays* ssp. *mays*) and squash (*Cucurbita* spp.) in Central New York. *American Antiquity* 72: 563–83.

Hart, J. P., R. A. Daniels, and C. J. Sheviak. 2004. Do *Cucurbita pepo* gourds float fish nets? *American Antiquity* 69: 141–48.

Hart, J. P., and W. A. Lovis. 2013. Reevaluating what we know about the histories of maize in northeastern North America: A review of current evidence. *Journal of Archaeological Research* 21: 175–216.

Haviland, W. A., and M. W. Power. 1981. *The Original Vermonters: Native Inhabitants, Past and Present*. Hanover, NH: University Press of New England.

Haynes, G. 2002. The catastrophic extinction of North American mammoths and mastodonts. *World Archaeology* 33: 391–416.

Helwig, K., J. Poulin, V. Monahan, and C. Thomas. 2021. Ancient throwing dart reveals first archaeological evidence of castoreum. *Journal of Archaeological Science: Reports* 37: 102949.

Hill, K. 1970. *More Glooscap Stories: Legends of the Wabanaki Indians*. New York: Dodd, Mead & Company.

Hubbard, L. L. 1879. *Summer Vacations at Moosehead Lake and Vicinity*. Boston: A. Williams & Company.

Hutchings, K. W. 2015. Finding the Paleoindian spearthrower: Quantitative evidence for mechanically-assisted propulsion of lithic armatures during the North American Paleoindian period. *Journal of Archaeological Science* 55: 34–41.

Isachsen, Y. W. 1981. Contemporary doming of the Adirondack Mountains: Further evidence from releveling. *Tectonophysics* 71: 95–96.

Jenkins, J., and A. Keal. 2004. *The Adirondack Atlas: A Geographic Portrait of the Adirondack Park*. Syracuse, NY: Syracuse University Press.

Jodry, M. 2005. Envisioning water transport technology in late Pleistocene America, pp. 133–60 in R. Bonnichsen, B. T. Lepper, D. Standord, and. M. Waters (eds.), *Paleoamerican Origins Beyond Clovis*. College Station: Texas A&M University Press.

Kaplan, L., T. F. Lynch and C. E. Smith Jr. 1973. Early cultivated beans (*Phaesolus vulgaris*) from an intermontane Peruvian valley. *Science* 179: 76–77.

Kelsay, I. T. 1984. *Joseph Brant 1743–1807: Man of Two Worlds*. Syracuse, NY: Syracuse University Press.

Kenny, M. 1995. *On Second Thought: A Compilation*. Norman: University of Oklahoma Press.

Kerber, J. E. (ed.) 2007. *Archaeology of the Iroquois. Selected Readings and Research Sources*. Syracuse, NY: Syracuse University Press.

Kimmerer, R. W. 2015. *Braiding Sweetgrass: Indigenous Wisdom, Scientific Knowledge and the Teachings of Plants*. Minneapolis: Milkweed Editions.

Kong, G. W. L. 2015. Fire histories during climate change in the Holocene at Wolf Lake, New York. Honours thesis for Bachelors of Science degree, Queens University, Kingston, Ontario.

Kuehn, S. R. 1998. New evidence for late Paleoindian–Early Archaic subsistence behavior in the western Great Lakes. *American Antiquity* 63: 457–76.

Kuhn, R. D. 2007. Reconstructing patterns of interaction and warfare between the Mohawk and Northern Iroquoians during the A.D. 1400–1700 period. Ch. 17 in J. E. Kerber (ed.), *Archaeology of the Iroquois: Selected Readings and Research Sources*. Syracuse, NY: Syracuse University Press.

LeBouef, K. A. 2014. Holocene vegetation, hydrology, and fire in the north-central Adirondacks of New York. Lehigh University Theses and Dissertations, Paper 1533. 70 pp.

Leeming, D. 2005. *The Oxford Companion to World Mythology*. New York: Oxford University Press.

Lemke, A. K. 2015. Great Lakes *Rangifer* and Paleoindians: Archaeological and paleontological caribou remains from Michigan. *Paleoamerica* 1: 276–83.

Lemke, A. K., and J. M. O'Shea. 2015. Hunters and hunting on the Alpena-Amberley Ridge during the Late Paleo-Indian and Early Archaic periods. Ch. 14 (pp. 169–76) in E. Sonnenburg, A. K. Lemke, and J. M. O'Shea (eds.), *Caribou Hunting in the Upper Great Lakes: Archaeological, Ethnographic, and Paleoenvironmental Perspectives*. Memoirs of the Museum of Anthropology, University of Michigan, No. 57.

Lewis, C. F. M., A. J. Breckenridge, and J. T. Teller. 2021. Reconstruction of isostatically adjusted paleo-strandlines along the southern margin of the Laurentide Ice Sheet in the Great Lakes, Lake Agassiz, and Champlain Sea basins. *Canadian Journal of Earth Science* 00: 1–21.

Loring, S. 1980. Paleo-Indian hunters and the Champlain Sea: A presumed association. *Man in the Northeast* 19: 15–38.

Lothrop, J. C., D. L. Lowery, A. E. Spiess, and C. J. Ellis. 2016. Early human settlement of northeastern North America. *PaleoAmerica* 2: 192–251.

Madrigal, T. C. 2017. "A Good Deal of an Archaeological Romance" A History of the Discovery and Excavation of the Lamoka Lake Site. Good-Deal-of-an-Archaeological-Romance_-History-of-the-Discovery-and-Excavation-of-the-Lamoka-Lake-Site_Madrigal_2017.pdf (lamokaledger.com).

Madrigal, T. C. 2000. Woodland period subsistence at Lamoka Lake: Animal bones from the Buffalo Museum of Science excavations. *The Bulletin, Journal of the New York State Archaeological Association* 116: 25–34.

Madrigal, T. C. 2001. Deer, passenger pigeons, and hunter-gatherers: Late Archaic subsistence and seasonality in central New York. *The Bulletin of the Archaeological Society of New Jersey* 56: 66–73.

Marcott, S. A., J. D. Shakun, P. U. Clark, A. C. Mix, et al. 2013. A reconstruction of regional and global temperature for the past 11,300 years. *Science* 339: 1198–1201.

Marschner, H., and O. K. Mason. 2013. The bow and arrow in northern North America. *Evolutionary Anthropology* 22: 133–38.

Mather, K. F. 1917. The Champlain Sea in the Lake Ontario basin. *Journal of Geology* 25: 542–54.

Mayewski, P. A., et al. 2004. Holocene climate variability. *Quaternary Research* 62: 243–55.

McAndrews, J. H., and C. L. Turton. 2010. Fungal spores record Iroquoian and Canadian agriculture in 2nd millennium A.D. sediment of Crawford Lake, Ontario, Canada. *Vegetational History and Archaeobotany* 19: 495–501.

McCauley, B., M. Collard, and D. Sandgathe. 2020. A cross-cultural survey of on-site fire use by recent hunter-gatherers: Implications for research on paleolithic pyrotechnology. *Journal of Paleolithic Archaeology*. https://doi.org/10.1007/s41982-020-00052-7.

Meltzer, D. J. 2009. *First Peoples in a New World: Colonizing Ice Age America*. Berkeley: University of California Press.

Monk, K. E. 1999. The development of aboriginal watercraft in the Great Lakes region. *Totem* 7, Article 9.

Noble, W. C. 2015 The Neutral Confederacy. *The Canadian Encyclopedia*. Accessed April 10, 2023 at https://www.thecanadianencyclopedia.ca/en/article/neutral.

Noren, A. J., P. R. Bierman, E. J. Steig, A. Lini, and J. Southon. 2002. Millennial-scale storminess variability in the northeastern United States during the Holocene epoch. *Nature* 419: 821–24.

O'Brien, M. J., et al. 2015. Transmission of cultural variants in the North American Paleolithic. Ch. 9 in A. Mesoudi and K. Aoki (eds.), *Learning Strategies and Cultural Evolution during the Palaeolithic*. Replacement of Neanderthals by Modern Humans Series. doi: 10.1007/978-4-431-55363-2_9.

Oswald, W. W., and D. R. Foster. 2011. Middle-Holocene dynamics of *Tsuga canadensis* (Eastern Hemlock) in northern New England, USA. *The Holocene* 22: 71–78.

Otis, M. 2018. *Rural Indigenousness: A History of Iroquoian and Algonquian Peoples of the Adirondacks*. Syracuse, NY: Syracuse University Press.

Overpeck, J. T. 1985. A pollen study of a late Quaternary peat bog, south-central Adirondack Mountains, New York. *Geological Society of America Bulletin* 96: 145–54.

Oyuela-Caycedo, A., and N. C. Kawa. 2020. A deep history of tobacco in lowland South America, pp. 27–44 in *The Master Plant*. doi: 10.4324/9781003087052-3.

Parker, A. C. 1910. Iroquois uses of maize and other food plants. New York State Museum Bulletin 144. University of the State of New York, Albany.

Parker, A. C. 1916. The origin of the Iroquois as suggested by their archaeology. *American Anthropologist* 18: 479–507.

Parker, R. L., et al. 2022. Laurentide Ice Sheet extent over the last 130 thousand years traced by the Pb isotope signature of weathering inputs to the Labrador Sea. *Quaternary Science Reviews* 287: 107564.

Pendergast, J. F. 1991. St. Lawrence Iroquoians: Their past, present, and immediate future. *Bulletin of the New York State Archaeological Association* 102: 47–74.

Piperno, D. R., A. J. Ranere, I. Holst, and R. Dickau. 2009. Starch grain and phytolith evidence for early ninth millennium B.P. maize from the Central Balsas River Valley, Mexico. *Proceedings of the National Academy of Sciences* 106: 5019–24.

Pohl, M. E. D., D. R. Piperno, K. O. Pope, and J. G. Jones. 2007. Microfossil evidence for pre-Columbian maize dispersals in the neotropics from San Andrés, Tabasco, Mexico. *Proceedings of the National Academy of Sciences* 104. doi/10.1073/pnas.07014.

Power, M. W., and J. B. Peterson. 1984. Seasons of prehistory: 4000 years at the Winooski site. Division for Historic Preservation, Montpelier, Vermont. Accessed January 10, 2023 at https://outside.vermont.gov/agency/ACCD/ACCD_Web_Docs/HP/Resources_Rules/Digging_Into_Archaeology/Seasons_of_Prehistory_4000_Years_at_the_Winooski_Site.pdf.

Prince, J. D. 1900. Some forgotten Indian place-names in the Adirondacks. *Journal of American Folklore* 13: 123–28.

Rafferty, S. M. 2006. Evidence of early tobacco in Northeastern North America? *Journal of Archaeological Science* 33: 453–58.

Rataul, R. C. 1996. New York state's garnetiferous gneiss bannerstones: The form and function of flight. *Northeast Anthropology* 72: 25–41.

Rayburn, J. A., D. A. Franzi, and P. L. K. Knuepfer. 2007. Evidence from the Lake Champlain Valley for a later onset of the Champlain Sea and implications for late glacial meltwater routing to the North Atlantic. *Palaeogeography, Palaeoclimatology, Palaeoecology* 246: 62–74.

Rayburn, J. A., P. L. K. Knuepfer, and D. A. Franzi. 2005. A series of large, Late Wisconsinan meltwater floods through the Champlain and Hudson Valleys, New York State, USA. *Quaternary Science Reviews* 24: 2410–19.

Rechlin, M., C. Laxson, C. Milewski, and C. Stager. 2019. Environmental history of Lower St. Regis Lake: Degradation and the path to ecological redemption. *Adirondack Journal of Environmental Studies* 23: 81–99.

Reichard, G. A. 1921. Literary types and dissemination of myths. *The Journal of American Folklore* 34: 269–307.

Reimer, P. J. 2021. Evolution of radiocarbon calibration. *Radiocarbon*. doi:10.1017/RDC.2021.62.

Ritchie, W. A. 1965. *The Archaeology of New York State*. Garden City, NY: The Natural History Press.

Robinson, F. J. 2017. Paleoindian sites, site patterning, and travel corridors along the southern arm of the Champlain Sea. Ch. 16 in J. A. M. Gingerich (ed.), *The Eastern Fluted Point Tradition*, Volume 2.

Robinson, F. J., IV, G. E. Springston, and J. G. Crock. 2017. Early and middle Paleoindian settlement patterns and the late-Pleistocene environment along the Champlain Sea. *PaleoAmerica*. doi: 10.1080/20555563.2017.1380997.

Robinson, F. W., IV. 2012. Between the mountains and the sea. Ch. 10 in C. Chapdelaine (ed.), *Late Pleistocene Archaeology & Ecology in the Far Northeast*. College Station: Texas A&M University Press.

Saint-Pierre, C.G., and R. G. Thompson. 2015. Phytolith evidence for the early presence of maize in southern Quebec. *American Antiquity* 80: 408–15.

Sanger, D. 1996. Gilman Falls site: Implications for the early and middle Archaic of the Maritime Peninsula. *Canadian Journal of Archaeology* 20: 7–28.

Sasso, J. 2023. The heart of the Adirondacks: The Totten & Crossfield Purchase. *Adirondack Almanac* (Feb. 10, 2023), accessed Feb. 1, 2024, at https://www.adirondackalmanack.com/2023/02/the-heart-of-the-adirondacks-the-totten-crossfield-purchase.html.

Schaefer, V. 2013. *Serendipity in Science: Twenty Years at Langmuir University*. Schenectady, NY: Square Circle Press, LLC.

Schneider, P. 1997. *The Adirondacks: A History of America's First Wilderness*. New York: Henry Holt and Company.

Shea, J. G. 1856. *Narrative of a Captivity Among the Mohawk Indians, a Description of New Netherlands in 1642–8, and Other Papers by Father Isaac Jogues, with a Memoir of the Holy Missionary by John Glimary Shea*. Press of the Historical Society, New York. Accessed April 10, 2023 at https://archive.org/details/cihm_39193/page/n4/mode/1up?ref=ol&view=theater.

Sherry, N. 2020. Early Adirondack surveys: The Great Corner and an ancient boundary. *New York Almanack* (Dec. 14, 2020). Accessed Feb. 1, 2024, at https://www.newyorkalmanack.com/2020/12/pioneer-adirondack-surveyor-a-great-discovery-an-ancient-boundary-line.

Sholts, S. B., J. A. M. Gingerich, S. Schlager, D. J. Stanford, and S. K. T. S. Wärmländer. 2017a. Time periods and morphological definitions for the styles of projectile points analyzed in this study. *PLoS ONE*. https://journals.plos.org/plosone/article?id=10.1371/journal.pone.0179933.

Sholts, S. B., J. A. M. Gingerich, S. Schlager, D. J. Stanford, and S. K. T. S. Wärmländer. 2017b. Tracing social interactions in Pleistocene North America via 3D model analysis of stone tool asymmetry. *PLoS ONE*. https://doi.org/10.1371/journal.pone.0179933.

Shuman, B. N., J. Marsicek, W. W. Oswald, and D. R. Foster. 2019. Predictable hydrological end ecological responses to Holocene North Atlantic variability. *Proceedings of the National Academy of Sciences* 116: 5985–90.

Smith, B. D. 1997. The initial domestication of *Cucurbita pepo* in the Americas 10,000 years ago. *Science* 276: 932–34.

Smith, B. D. 2006. Eastern North America as an independent center of plant domestication. *Proceedings of the National Academy of Sciences* 103: 12223–28.

Smith, B. D., and R. A. Yarnell. 2009. Initial formation of an indigenous crop complex in eastern North America at 3800 B.P. *Proceedings of the National Academy of Sciences* 106: 6561–66.

Snow, D. 1998. *The Iroquois*. Oxford, UK: Blackwell.

Sonnenburg, E., A. K. Lemke, and J. M. O'Shea (eds.). 2015. *Caribou Hunting in the Upper Great Lakes: Archaeological, Ethnographic, and Paleoenvironmental Perspectives*. Memoirs of the Museum of Anthropology, University of Michigan, No. 57.

Sorenson, A. 2019. The uncertain history of fire-making by humans: The state of the art and smoldering questions. *Mitteilungen der Gesellschaft für Urgeschichte* 28: 11–50.

Speck, F. G. 1923. Algonkian influences upon Iroquois social organization. *American Anthropologist* 25: 219–27.

Speck, F. G. 1940. *Penobscot Man*. Philadelphia: University of Pennsylvania Press.

Speth, J. D., K. Newlander, A. A. White, A. K. Lemke, and L. E. Anderson. 2013. Early Paleoindian big-game hunting in North America: Provisioning or politics? *Quaternary International* 285: 111–39.

Spiess, A. E., B. J. Bourque, and R. M. Gramly R.M. 1984. Early and middle Archaic site distribution in western Maine. *North American Archaeologist* 4: 225–44.

Stager, J. C., B. F. Cumming, K. Laird, A. Garrigan-Piela, N. Pederson, B. Wiltse, C. S. Lane, J. Nester, and A. Ruzmaikin. 2016. A 1,600-year record of hydroclimate variability from Wolf Lake, NY. *The Holocene*. doi: 10.1177/0959683616658527.

Stager, J. C., D. Fadden, and C. B. Wolff. 2022. Dugout canoes from lakes of the Adirondack uplands. *Adirondack Journal of Environmental Studies* 25, Article 3. https://digitalworks.union.edu/ajes/vol25/iss1/3/.

Stager, J. C., B. Wiltse, B. F. Cumming, T. C. Messner, J. Robtoy, and S. Cushing. 2021. Hydroclimatic and cultural instability in northeastern North America during the last millennium. *PLoS ONE* 16: e0248060. https://journals.plos.org/plosone/article?id=10.1371/journal.pone.0248060.

Steadman, D. W., T. W. Stafford, and R. E. Funk. 1997. Non-association of Paleoindians with AMS-dated late Pleistocene mammals from the Dutchess Quarry Caves, New York. *Quaternary Research* 47: 105–16.

Stuiver, M., P. J. Reimer, and R. W. Reimer. 2022. CALIB 8.2 program. Available at http://calib.org.

Sulavik, S. B. 2007. *Adirondacks: Of Indians and Mountains, 1535–1838*. Fleischmanns, NY: Purple Mountain Press.

Taché, K., and O. E. Craig. 2015. Cooperative harvesting of aquatic resources and the beginning of pottery production in north-eastern North America. *Antiquity* 89: 177–90.

Taché, K., D. White, and S. Seelen. 2008. Potential functions of Vinette I pottery: Complementary use of archaeological and pyrolysis GC/MC data. *Archaeology of Eastern North America* 36: 63–90.

Talbot, F. X. 1935. *Saint Among Savages: The Life of Isaac Jogues*. New York: Harper & Brothers.

Taylor, W. B. 2009. Fire-making kits. *Bulletin of the Massachusetts Archaeological Society* 70: 3–6.

Tehanetorens (Ray Fadden). 1998. *Legends of the Iroquois: As Told by Tehanetorens (Ray Fadden)*. Summertown, TN: Book Publishing Company.

Terrie, P. G. 1997. *Contested Terrain: A New History of Nature and People in the Adirondacks*. Syracuse, NY: Syracuse University Press.

Thomas, K. A., B. A. Story, M. I. Eren, B. Buchanan, B. N. Andrews, M. J. O'Brien, and D. J. Meltzer. 2017. Explaining the origin of fluting in North American Pleistocene weaponry. *Journal of Archaeological Science* 81: 23–30.

Tobiessen, P. 2012. *The Secret Life of a Lake: The Ecology of Northern Lakes and Their Stewardship*. Niskayuna, NY: Graphite Press.

Tremblay, M., J. C. Stager, J.-M. St-Jacques, S. Murphy, M. Peros, and B. S. Carl (in press, 2024). A 13,000-year history of vegetation and fire in a rare inland pine barrens: The Albany Pine Bush (Albany County, New York, USA). *PLoS ONE*, doi .org/10.1371/journal.pone.0314101.

Tushingham, S., C. M. Snyder, K. J. Brownstein, W. J. Damitio, and D. R. Gang. 2018. Biomolecular archaeology reveals ancient origins of indigenous tobacco smoking in North American Plateau. *Proceedings of the National Academy of Sciences* 115: 11742–47.

Van Diver, B. B. 1985. *Roadside Geology of New York*. Missoula, MT: Mountain Press.

Wales, N., et al. 2019. Ancient DNA reveals the timing and persistence of organellar genetic bottlenecks over 3,000 years of sunflower domestication and improvement. *Evolutionary Applications* 12: 38–53.

Wardner, C. A. 2010. *Footprints and Sunset on Adirondack Trails: The Memoirs of James Manchester Wardner, 1831–1904*. Jay, NY: Graphics North.

Webster, D. A. (no date) Early history of Atlantic salmon in New York. Accessed January 5, 2023, at http://www.fishcreeksalmon.org/history-atlantic-salmon.htm.

Wheeler, R. J., et al. 2003. Archaic period canoes from Newnan's Lake, Florida. *American Antiquity* 68: 533–51.

Whitehead, D. R., and S. T. Jackson. 1990. The regional vegetational history of the High Peaks (Adirondack Mountains), New York. New York State Museum Bulletin No. 478. 27 pp.

Whittaker, J. C., D. B. Pettigrew, and R. J. Grohsmeyer. 2017. Atlatl dart velocity: Accurate measurements and implications for Paleoindian and Archaic archaeology. *PaleoAmerica* 3: 161–81.

Whyte, T. R. 2007. Proto-Iroquoian divergence in the late Archaic-early Woodland period transition of the Appalachian highlands. *Southeastern Archaeology* 26: 134–44.

Wiseman, F. M. 2001. *The Voice of the Dawn: An Autohistory of the Abenaki Nation*. Lebanon, NH: University Press of New England.

Yansa, C. H., and K. M. Adams. 2012. Mastodons and mammoths in the Great Lakes region, USA and Canada: New insights into their diets as they neared extinction. *Geography Compass*: 1–14. doi 10.1111/j. 1749-8198.2012.00483.x.

Younging, G. 2018. *Elements of Indigenous Style: A Guide for Writing By and About Indigenous Peoples*. Brush Education, Inc., Edmonton, Alberta, Canada.

Zazula, G. D., et al. 2014. American mastodon extirpation in the Arctic and Subarctic predates human colonization and terminal Pleistocene climate change. *Proceedings of the National Academy of Sciences* 111: 18460–65.

Zhang, W., et al., 2022. Holocene seasonal temperature evolution and spatial variability over the Northern Hemisphere landmass. *Nature Communications* 13: 5334–46.

ACKNOWLEDGMENTS

Many knowledgeable people have generously contributed their expertise to this project, and we cannot thank them enough. Those contributors include, in alphabetical order, Jim Bickford, Jim Blackmore, Sally Bogdanovitch, Darren Bonaparte, Jerry Bottcher, Brian Carl, Mateo Champlain, Tsiorasa Cook-Barreiro, Justin Cree, Brian Cumming, Don Fadden, Elizabeth Karonhisake Fadden, John Kahionhes Fadden, David Franzi, Albert Fulton, Brian Grills, John Hart, Sue Herne, Claire Horn, Steven Jackson, Margaret Lafrance, Corey Laxson, John Lothrop, Tim Messner, Laurie Miroff, Melissa Otis, Neil Patterson, Ralph Rataul, Francis "Jess" Robinson, Daniel Seib, Jay Stager, Paul-René Tamburro, Jim Tucker, Susan Winchell-Sweeney, and Chris Wolff. Asha Stager made valuable editorial suggestions for early drafts of the manuscript, and Ken Alton, Alex Comer, Bethany Garretson, and Kary Johnson provided valuable assistance in the field. Sarah Galvin of The Bookstore Plus connected us to North Country Books and Globe Pequot by way of Dennis Hayes. Financial support for this and related research was provided to Curt Stager by Caroline Draper-Lussi through the Draper-Lussi Endowed Chair fund at Paul Smith's College and by P2C2 grants from the National Science Foundation. Our thanks and sincere apologies to anyone whom we might have accidentally left off the list, as well!